SOCIAL IMAGING

Social Capital and Spatial Use Patterns
with Potential Correlations

by

Milton J. Friesen

St. Catharines, ON, Canada

Summary

Social capital – understood as a complex set of phenomena characterized by human interactions involving trust and social ties – is a significant common good resource that is necessary for community and city well-being. This resource can vary in quantity and type depending on context. Within the range of phenomena that comprise social capital, trust (individual and institutional) and social networks (friends, relatives, and acquaintances) are critical and will be examined in what follows.

Our empirical understanding of social capital dynamics is incomplete including consistent and cost-effective social capital measurements at neighbourhood levels. Exploring and establishing correlations between social capital levels and more readily obtained GPS spatial use data may provide an efficient proxy for measuring social capital.

This possible correlation is examined by looking at the formal relationship between social capital levels as collected by a survey instrument and the spatial patterns produced by human movement in an urban setting. Visualizing social capital measurement would assist urban and social planners in understanding an important set of features reflective of human well-being. This would complement other health and well-being measures currently being used.

The results of my examination of both social capital and movement patterns for a sample of citizens in Hamilton, ON (n=97) suggests that no simple proxy measure was identified but that some spatial measures, such as total distance travelled and the size of the range for individuals, have potential for further investigation. Linear regression, random forest and principal component techniques were used to explore these potential correlations. The concluding chapter makes use of Christopher Alexander's work to argue for continuous linking of the social and spatial in policy and design processes.

Data sharing is not possible given the data collection agreement that individual participant data results would only be communicated in aggregate. The methodology and instrument developed alongside continuously improving, lower cost GPS data and spatial device data offer exciting opportunities for scalable social capital data visualization.

Acknowledgements

This work has been patiently supported by my wife, Michelle, and our children (Mackenna, Dylan, Brittany, and Charae) as well as by a host of extended family, friends, and colleagues over many years. The East Hamilton participants who agreed to complete surveys and carry a GPS data collection device with them for a week were critical in testing the ideas I set out explore. Without them there would be no primary data to analyze.

I would also like to thank the School of Planning (University of Waterloo) faculty who served as advisors, committee members and informal supporters. Clarence Woudsma was the Director during the years I was there and offered opportunities to develop and teach a range of undergraduate and graduate courses which were a highlight. Dawn Parker and Rob Feick were my advisors and Pierre Filion, Jennifer Dean, and Shreyas Sundaram (now at Purdue) served as dissertation committee members. After nearly five years of trying, we could not come to an agreement on the final form for this dissertation. I acknowledge the ways in which their interactions promoted a refinement of important elements. Exploratory work is rarely a neat process and the range of motion we can each tolerate is varied enough to make collective exploration difficult. My goal for this advanced graduate work was to maximize learning. As that curve flattened, I decided it was time to engage other pursuits while remaining grateful for what was captured in the research presented here. I hope to have made good mistakes in what follows.

Author Bio

Milton Friesen has experience in municipal government, policy, advanced academic study, and marketplace development. His research has been focused on exploring the social infrastructure of communities and cities. He combines practical, direct experience with leading edge methods in urban data acquisition, analysis, and application to pressing contemporary issues. He has served on the steering committee of the Thriving Cities Project at the Institute for Advanced Studies in Culture (University of Virginia), has been a member of the Program Committee for the Computational Social Sciences Society of the Americas, and pursued graduate studies at the University of Waterloo (School of Planning). He currently serves as the Managing Director of CitiIQ, a global measurement platform for urban monitoring.

The hallmark of good science is that it uses models and 'theory' but never believes them.
--Martin Wilk

Far better an approximate answer to the right question, which is often vague, than an exact answer to the wrong question, which can always be made precise.
--John Tukey "The Future of Data Analysis" 1962

[Utake] Taniyama was not a very careful person as a mathematician. He made a lot of mistakes but he made mistakes in a good direction. So eventually he got to right answers. I tried to imitate him, but I found it is very difficult to make good mistakes.
--Prof. Goro Shimura, Princeton University (A close friend of Taniyama in Japan at the time of his very unexpected death)

Table of Contents

List of Figures ix

List of Tables xiii

Chapter 1: Planning and the Measurement of Social Capital 14

1.1 Introduction 14

1.2 Central Hypothesis 16

1.3 Secondary Hypothesis 16

1.4 Research Questions 17

1.5 Definitions of Social Capital 18

1.6 Planning and Context for Social Capital 19

1.7 History of Social Capital 25

1.7.1 Current Academic Research on Social Capital 28
1.7.2 Research by Discipline 28
1.7.3 Research Across Time 32
1.7.4 Research of Abstract Themes 33
1.7.5 Significance of Social Capital Research Overview 36

1.8 Social Capital Research from Philosophy to Application 38

1.9 Methodology for Social Capital Research 42

1.9.1 Social Capital Measurement Frameworks 44
1.9.2 Planning and the Measurement of Social Capital 45
1.9.3 Issues of Scale in Social Capital Measurement 45
1.9.4 Selecting a Social Capital Measurement Tool 46
1.9.5 Social Capital and other Social Phenomena 46

1.10 Significance for Planning Policy 47

1.11 Conclusion 48

Chapter 2: Advancing Social Capital Measurement: Using General Social Survey Cycles to Develop an Efficient Survey Instrument 51

2.1 Introduction 51

2.2 Social Capital Measurement Approaches 51

2.3 Survey Design **52**

2.4 Comparing Survey Data with Spatial Data **53**

2.5 The Nature of GSS Data **54**

2.5.1 Selecting Questions from Existing GSS Cycles 55
2.5.2 Identification Codes for SCGSS 41 Questions 60

2.6 Testing SCGSS **64**

2.7 Deployment of SCGSS **65**

2.7.1 Identification of Census Tracts for comparison 65
2.7.2 Recruitment of participants 70
2.7.3 Data Collection 73
2.7.4 Data Handling Process 74
2.7.5 Geographic Distribution of Completed Surveys 75

2.8 Results of SCGSS **76**

2.8.1 Context Analysis 79
2.8.2 Social Network Characteristics 82
2.8.3 Trust 89
2.8.4 Synthesis of Social Capital Measures – Networks and Trust 97

2.9 Results of Synthetic Social Capital **99**

2.9.1 Social Network Size 99
2.9.2 Trust 101

2.10 Discussion **106**

CHAPTER 3: Ten Exploratory Metrics for Comparing GPS Spatial Use Patterns with Socio Economic Data in Urban Settings **108**

3.1 Introduction **108**

3.2 Methods for Spatial observation through GPS data logging **108**

3.3 Time Use **110**

3.4 Data Collection **111**

3.5 Spatial Data Processing **114**

3.6 Spatial Variables Selected for Analysis **116**

3.7 Results of Spatial Analysis **129**

3.8 Modal Analysis **136**

3.9 Discussion of Demographic and Spatial Comparison 138

Chapter 4: An Exploratory Approach for Measuring Social Capital by using Spatial Data 143

4.1 Introduction 143

4.2 Social Capital Data 146

4.3 Spatial Data 148

4.4 Methods for Comparing Spatial and Social Capital Data 148

4.4.1 Linear Model 150
4.4.2 Random Forest Model 151
4.4.3 Principal Component Analysis 152

4.5 Results 153

4.5.1 Summary of Linear Model Results 154
4.5.2 Summary of Random Forest Model Results 157
4.5.3 Summary of Principal Component Analysis Results 159
4.5.4 Comparison of Linear, Random Forest and Principal Component 163

4.6 Discussion 166

4.7 Conclusion 170

Chapter 5: Implications of Spatial Planning and Social Infrastructure 172

5.1 Introduction 172

5.2 Design Insight Through Social and Spatial Integration 174

5.3 Fifteen Transformations into Five Questions for Interrogative Planning 176

5.3.1 Is it subjective? 177
5.3.2 Does it cohere? 179
5.3.3 Is it adaptive? 181
5.3.4 Is it recursive? 183
5.3.5 Is it a functional whole? 185

5.4 Disciplines for Integrating Social and Spatial Research and Practice 186

5.4.1 Network Theory and Mathematical Capability 187
5.4.2 Adaptive and Effective Physical Infrastructure 188
5.4.3 Improved Data Sources and Structures 189

5.5 Conclusion: The Future of Planning as a Synthesis of the Social and the Spatial 191

References......189

List of Figures

Figure 1-1 - WoS™ Research Categories by number of records. 29

Figure 1-2 - WoS™ social capital coloured citation graph by categories. Red is medicine, orange is Psychology, light green (lower section) is biology, brown is Earth, Environmental, Engineering, yellow at top is sociology, lightest green (top) is planning and public administration, etc. Citation data was extracted from WoS™, converted to .gexf format using an Excel macro and then imported into Gephi for visualization and analysis. Square is area highlighted in the following (Figure 1-3)...... 30

Figure 1-3 - Magnification of a selection of the WoS™ social capital coloured citation graph by categories. 31

Figure 1-4 - 100 citations from top 500 most cited (limited to 100 nodes for clarity). Groups formed by citation closeness. In this case, the oldest paper is at the top Banfield (1958) and the newest is at the bottom Ellison (2007). Data downloaded from Web of Science August 29, 2019. 32

Figure 1-5 - Publication of social capital papers by year. 33

Figure 1-6 - Web of Science social capital abstracts with words occurring more than 5000 times arranged by frequency. [Raw text was processed in the following way: punctuation removed, numbers removed, converted to lower case, stopwords removed using "english" stopwords from R (n=174), removed definition words ("social", "capital") whitespace removed, and frequency matrix generated for analysis]. 35

Figure 1-7 - Word frequency for abstracts that contain both "social capital" and "spatial" as search terms for WoS. 36

Figure 1-8 - Progressive framework for social capital from philosophy to the application. 38

Figure 2-1 - *Logic diagram representing the process used to reduce 27 GSS cycles into the Social Capital GSS (SCGSS) instrument.* 56

Figure 2-2 - Comparison of selected GSS modules (lower nodes) with the PSCS survey control instrument (top nodes - Q1-8 bonding capital and Q9-16 bridging capital) using social capital themes of trust and social networks. Graphic drawn by author. 59

Figure 2-3: Summary statistics and histogram of income distribution for 130 Hamilton Ontario Census Tracts. Data from Statistics Canada, 2011. Histogram represents average income (ages 18-64) and number of Census Tracts in that income range...... 66

Figure 2-4: The white circle in the top image is a 2km radius around the center point of the three core Census Tracts. The lower map denotes the primary boundaries of the study area: Ottawa Street to the West, Barton Street to the North, the Red Hill Expressway to the East and the Escarpment to the South. 68

Figure 2-5 - Historical map of 1930 transit routes with "Bartonville Bus" and dashed city boundary line showing Rosedale as an undeveloped area (Hamilton Public Library Archives...... 70

Figure 2-6 - Core Census Tracts in solid colours, secondary tracts share common colour but lighter. Priority established by common income averages + proximity. Census tract boundaries and fill done with ArcGIS. Basemap provided by Open StreetMap (and) contributors. 73

Figure 2-7 - Marital status across three Census Tracts (weighted by count). 76

Figure 2-8 - Google Earth images of Census Tracts reflecting basic urban forms. Google Earth tours of each of the three core Census Tracts are also available. 77

Figure 2-9 - Income distributions by Census Tract (weighted by count). Y axis = number of participants in each category. Missing bars reflect no participants in that particular bracket. 78

Figure 2-10 - Age across three Census Tracts (weighted by count). 79

Figure 2-11 - Comparison of general neighbourhood network across all three Census Tracts (weighted by count). 80

Figure 2-12 - Responses to question about how much respondent felt they belonged in their community (weighted by count). 81

Figure 2-13 - Comparison of three Census Tracts and perceptions of charitable donations and volunteering (weighted by count). 82

Figure 2-14 - Comparison of social network size across the three Census Tracts (weighted by count). 83

Figure 2-15 - Modes of contact for various social groups across all three Census Tracts (weighted by count). 85

Figure 2-16 - Participants feelings about the frequency of contact with friends, family and acquaintances compared across all three Census Tracts (weighted by count). 87

Figure 2-17 - Walking distance comparison across all three Census Tracts (weighted by count). 89

Figure 2-18 - Trust across all relationships and trust in a specific, unknown person or persons across all three Census Tracts (weighted by count). The absence of "Low Trust" and "High Trust" for "Trust People in General" is owing to missing values in the survey instrument. These categories for trust are taken from the Canadian General Social Survey instrument noted earlier. 90

Figure 2-19 - Trust across three distinct community grouping spaces for all three Census Tracts (weighted by count). 92

Figure 2-20 - Trust in core institutions - police, schools, and city officials (weighted by count). 94

Figure 2-21 - Trust in local religious, local business and big business (weighted by count). 95

Figure 2-22 - Trust in healthcare and welfare (weighted by count). 96

Figure 2-23 - Graph of social network sizes aggregated by respective study areas in both bar graph form with social sections retained (above) and via a trend line where social sections are aggregated. 100

Figure 2-24 - Trust Quotients compared across Delta, Rosedale and McQueston (weighted by count). Trust quotients calculation: Likert score (1 is low – 5 is high) is multiplied by number of participants in each bracket. Individual trust measures occur earlier in this chapter. 104

Figure 2-25: Trust Quotients by individual Census Tract (weighted by count) and individual questions combined from earlier individual graphs. 105

Figure 2-26 - Aggregate of all Trust scores. Red dashed upper line represents maximum amount of trust where all responses are total trust (5) on the Likert scale – an upper limit based on the survey value options. 106

Figure 3-1 - Participant responses to the normalcy of the week in which they carried the GPS data logging devices (all participants)........ 110

Figure 3-2 - Comparison of three study areas in terms of time use blocks (weighted by number of participants in each Census Tract) - Delta(High), Rosedale(Mid), McQueston(Low). The x-axis represents the number of 6 hour blocks designated for each primary activity. 111

Figure 3-3 - Time based distribution of data collection periods for each of the 97 participants (y axis = privacy ID for each participant). For example, the lower left participant AT25N began GPS data logging on January 20, 2016 and ended (4) days later on January 25. 113

Figure 3-4: Calculating the Mean Center for a set of geographic coordinates. Figures are modifications from ESRI technical workshop graphics (Bennett and Vale, 2015). 119

Figure 3-5 - Standard distance from the mean - radius of circle around mean centre of GPS features represents one standard deviation. Figures are modifications from ESRI technical workshop graphics (Bennett and Vale, 2015). 120

Figure 3-6 - Median value (green) for a given set of geographic features. In this example with ten features, the fifth value is the median (upper element). If the line is stretched out, the logic of the process becomes clear (lower element). Author designed graphic. 121

Figure 3-7 - Directional orientation, or linear mean direction, of a given set of features. Figures are modifications from ESRI technical workshop graphics (Bennett and Vale, 2015)........ 122

Figure 3-8 - Examples of XY ratio of ellipse measurements. 123

Figure 3-9 - Determination of ellipse through calculation of x and y-axis standard deviations from the mean. Figures are modifications from ESRI technical workshop graphics (Bennett and Vale, 2015)......... 125

Figure 3-10 - Nearest neighbour conceptual description. Figures are modifications from ESRI technical workshop graphics (Bennett and Vale, 2015). 126

Figure 3-11: Total area of a rectangle covering all features in a given dataset........ 128

Figure 3-12 - Rectangle enclosing all points used to calculate total area of travel territory........ 128

Figure 3-13 - Comparison of total distance of GPS track measured by calculating distance between each reading (unweighted). 129

Figure 3-14 - Size of circle drawn around the mean centre determined by standard deviation from the mean........ 130

Figure 3-15 - Comparison of three Census Tracts using the difference between mean center and median center........ 130

Figure 3-16 - Deviational Ellipse Rotation. Comparison of the degree of rotation of the standard deviational ellipse across three Census Tracts(degrees clockwise from North). Rotation is 3 degrees more easterly in the middle income neighbourhood than in the Delta(High). The lower income neighbourhood is 12 degrees more easterly than the Delta(High). 131

Figure 3-17: Ratio of the X and Y axes showing relative degree of roundness 132

Figure 3-18 - Average mean distance between GPS data reading points. 133

Figure 3-19 - Comparison of Z-scores across three Census Tracts. 133

Figure 3-20: Comparing degree of randomness in point to point ratios across three Census Tracts. 134

Figure 3-21 - Area of rectangle drawn around all GPS data points - mean average of all tracks per Census Tract. 135

Figure 3-22 - Standard distribution of the mean value of "speed" variable across all GPS data points by Census Tract. 135

Figure 3-23 - Results of modal analysis of social imaging data by neighbourhood. "None" in both bar graphs and pie charts means no movement. 137

Figure 3-24 - Correlation table for Standard Distance Radius and Demographic variables. 138

Figure 3-25 - Correlation table for Total Distance and Demographic variables 139

Figure 3-26 - Correlation table for Difference of Mean / Median Centre and Demographic variables 139

Figure 3-27 - Correlation table for Ellipsis Rotation and Demographic variables 140

Figure 3-28 - Correlation table for Ellipsis X and Y Axes Ratio and Demographic variables 140

Figure 3-29 - Correlation tables for Nearest Neighbor Ratio and Nearest Neighbor Z-Score and Demographic variables 141

Figure 3-30 - Correlation table for All Points Area Rectangle and Demographic variables 141

Figure 4-1 - Diagram of SCGSS social capital survey instrument question themes (demographic, social network and trust) along with an image of a GPS data track that comprises the spatial measures. 147

Figure 4-2 - Random forest algorithm logic. For variable selection, it should be noted that the rfm() model in R determines the number of variables chosen by using the square root of total variables for classification random forests (i.e. dependent variable is a factor) and the number of total variables divided by 3 for regression random forests (i.e. dependent variable is continuous). The number of variables can be specified manually. 152

Figure 4-3: Total synthetic social capital value (all.socap) distribution across all 97 participants. 154

Figure 4-4 - Principal component analysis for **all.socap** (PCA1 model - total social capital) variables with the circle indicating the spread of 50% of the data points. 161

Figure 4-5: Principal component analysis for sns.socap (PCA2 model - social network related social capital) variables with the circle indicating the spread of 50% of the data points. 162

Figure 4-6: Principal component analysis for **ts.socap** (PCA3 model - trust related social capital) variables with the circle indicating the spread of 50% of the data points. 163

Figure 5-1 - Alexander's 15 transformations can be usefully synthesized into 5 orienting questions. The transformations, synthesis and orienting questions can be used as a framework for model specification and design considerations depending on context. 176

List of Tables

Table 1-1 Selected social capital scholarship highlights spanning 153 years (book titles in italics). 26

Table 1-2 Summary of WoS™ social capital textual data with "spatial" subset. 34

Table 2-1: Overview of Canadian GSS cycles with dates and themes ("CanSIM General Social Survey Overview," 2015). 53

Table 2-2: Final SCGSS Question Categories.. 60

Table 2-3: Census Tract identification without regard for context. 67

Table 2-4: Context adjusted Census Tracts. 67

Table 2-5 *Data utilized by Malatest (survey firm) Methodology Report. Population boundary is adults ages 18-64 who live in Census Tract. Location confirmation required a landline where name and address could be correlated. Confidence interval at 95%. ASDE Survey.* 71

Table 2-6 *Secondary Census Tract options in order of importance within each neighbourhood.* 72

Table 2-7 *Weighted income distributions across all Census Tracts where "Income" question was not refused.* 78

Table 2-8 *Summary table of synthetic capital variables. Variables formatted for use in R. Some values have a minimum of "5" where the lowest value choice on the survey was a 1 rather than a 0. *Original Likert values in the GSS required that these values be changed from ascending to descending before including. **This allows the use of* **all.socap** *as a dependent variable to compare against the remaining independent variables – Q1-16 and the 10 Spatial Variables*... 98

Table 2-9 Social network sizes (number of people each participant in each of the *close friends, relatives,* and *acquaintances* categories) and totals for each of the respective study areas (weighted). 101

Table 2-10 *Aggregate Trust quotients as totals for people overall and institutions overall. The Trust Quotient is a total of both category aggregates (weighted).* 103

Table 3-1 *Raw data format as downloaded from BT-Q1000 X devices.* 114

Table 3-2 Variables used for spatial analysis – data table (variable names) as used in R-based coding of models and data frames. Detailed descriptions of variables are available (Ebdon, 1991; Mitchell, 1999, 2005; L. Scott & Janikas, 2010). *Synthetic variables developed for this paper. 117

Table 4-1 Initial and tuned model results summary.. 155

Table 4-2 Results of random forest analysis for central and secondary hypotheses. 158

Table 4-3 Principal component analysis results summarized.. 160

Table 4-4 *Summary and comparison of linear model, random forest and principal component analyses.* 164

Chapter 1: Planning and the Measurement of Social Capital

1.1 Introduction

Social capital consists in the relational ties between people. These ties form patterns and dynamics that can be partially measured in spatial terms – time, location and duration. Perceptions of other people – whether we trust them or not and in what way – are constantly changing as we move among colleagues, friends, family members and neighbours. These relational webs of social capital have many dimensions, exhibit emergent behaviour, and are constantly changing (Trofimova, 2000). The dynamic nature of human connectivity means that individuals both direct (individual agency) and are directed by (structural determinism) those interactions (Gunderson & Holling, 2002; Ried et al., 2019) in ways that impact individuals and communities.

Although they are invisible, relational resources at individual and institutional scales are very powerful and profoundly shape our lives (J. S. Coleman, 1988; Hanifan, 1916; N. Lin et al., 2001; Ostrom & Ahn, 2003; R. D. Putnam, 2001; Tocqueville, 2001). This power is expressed through the dynamics of our communities and cities, sometimes driving tectonic changes in urban settings. Advancing our understanding of these dynamics across scales is growing more urgent as both formal and informal urbanization intensifies.

For example, scholarship that suggests low levels of social capital are linked to chronic poverty (Goldstein et al., 2009), inability to resist negative change (Jacobs, 1959) and an inability to respond successfully to natural and ecological disasters (Burdett & Sudjic, 2011, p. 83) with the inverse characterizing areas where social capital is high. Social isolation (which reflects low levels of social capital) has increased to the point that it is now identified as a significant health risk on par with heart disease and smoking (Holt-Lunstad et al., 2010). Other well-being research in developed countries supports the conclusion that increased social isolation leads to decreased health and wellbeing (Berkman et al., 2004; Eng et al., 2002; Forster & Stoller, 1992; Giles et al., 2005; Horsten et al., 2000; Kawachi et al., 1996; Pantell et al., 2013; Schoenbach et al., 1986; Vezina, 2011; Yasuda et al., 1997; X. Zhang et al., 2007).

In this book, social capital is defined as *the density and quality of human relational networks both formal and informal*. This includes the number and type of relationships an individual has and the degree of trust present in both those relationships and in the wider contexts beyond personal ties.

Social capital as a concept has received increased attention over the last two decades. Researchers have sought to define what social capital is, what it does and improve measurement of its dynamic properties through development of relevant statistical data, social capital indexes, social capital modules in census projects, meta-analysis of social capital and a wide variety of social and spatial network structure and process techniques (Adrienko & Adrienko, 2011a; Bivand, 2013; Ester et al., 1996; Farber et al., 2014; Franke, 2005; Neutens et al., 2013). Measuring social capital more easily and accurately would support more reliable and consistent social impact measurements for planners specifically and for the NGO, government, and private philanthropic sectors more generally (Antadze & Westley, 2012; M. Granovetter, 2005; L. Nguyen et al., 2015).

Despite advances in technology that make it easier for people to interact over large distances, spatial location persists as an important feature of our closest and most developed social ties (Onnela et al., 2011). This spatial aspect of relationships may provide a key for accessing better social capital measurement when paired with our ability to very accurately measure the spatial movement patterns of individuals using GPS data logging technology (L. Zhang et al., 2011). Linking spatial and social data in formal ways will support our understanding of the nature of correlations between movement data and social capital levels. Even if such a correlation was partial, it could be useful in the "necessarily approximate nature of useful results in data analysis" (Tukey, 1962). A particularly important benefit of such research would be the ability to capture more accurate social capital levels at local or neighbourhood scales where the consequences of relational networks have the greatest significance and potentially fill in the gap left by national scale government research tools such as the General Social Survey which lack local scale resolution.

I propose that one means of improving social capital measurement is to compare social capital levels with the movement patterns of individuals. A spatially informed proxy measure could dramatically improve our ability to see the invisible social structural fabric of the city and thus to better understand the social impact of planning activity more clearly. The challenge is that the variety and nature of human social interactions in high density locations like cities are extremely complex. New synthetic research design and analysis drawing on analytical sociology, computational social science, machine learning and spatial analytics will support my argument and deepen our insight into these social dynamics (Dandekar, 2005; Keuschnigg et al., 2018) offering possibilities for further development of improved methods.

Many aspects of social capital are experienced locally, by individuals in neighbourhoods, rather than abstractly at national levels. New social science methods and techniques are emerging that allow us to understand these dynamics with increasing. Labour intensive interviews and data gathering have made social network research difficult. New forms of observation driven data collection, such as GPS data logging combined with traditional tools like surveys, enable growing knowledge of the nature of social capital people in a given area. Planners can use this insight to more intelligently direct limited resources and specific planning solutions more effectively (Aaltonon et al., 2005). This book examines the nature of social and spatial interactions at a neighbourhood scale by testing the following hypotheses.

1.2 Central Hypothesis

The primary purpose of the following examination is to advance social capital research by exploring potential correlations between movement patterns and social capital levels. The **central hypothesis** is that *social capital levels decrease with decreasing spatial movements* (Aaltonon et al., 2005; Adrienko & Adrienko, 2011b; Brockmann et al., 2006; Crandall et al., 2010; Diaconis & Mosteller, 1989; González et al., 2008; Mastrandrea et al., 2015; Neutens et al., 2013; Sekara et al., 2016). This is based on the working logic that an individual who never leaves their home will have fewer in-person social contacts, lower levels of direct community institutional interaction, and as a result, lower levels of social capital than an individual who is more active through work, school, shopping, recreation, and other human activities.

1.3 Secondary Hypothesis

A **secondary hypothesis** asserts that social capital levels do not follow income levels and that the dynamics of social capital are more complex than a simple correlation with income would suggest. Social resource levels can be low even among people with higher incomes, and social isolation appears to be present across income groups (Lindström & Lindström, 2006; Macinko & Starfield, 2001; Pearce & Davey Smith, 2003). The exploratory nature of this research suggests that it is important to include income inequality as an important possible source of variation in social capital levels.

While testing these hypotheses, a range of research questions will be utilized to focus specifically on capturing relevant aspects of movement, trust and social networks for individuals across three specific local communities.

1.4 Research Questions

We need strong guidelines to determine how to approach a difficult problem such as the one proposed above. Research questions are an indispensable aid in that process and those questions are divided into two types: **primary** research questions and **secondary** research questions.

Primary research question: *What is the relationship between how a statistically significant group of people move around their neighbourhoods, reflected in GPS spatial data, and social capital levels as established by the SCGSS baseline measurements?*

Secondary research questions:

1. *How can we improve our ability to measure social capital at a local level (e.g. Census Tract)? What are the challenges or limits to undertaking this scale of measurement?*

2. *How could the Canadian General Social Survey (GSS) questions be used to measure baseline social capital levels at sub-regional levels?*

3. *Are there relationships between how a sample of individuals move around their neighbourhoods as reflected in GPS spatial data, and social capital levels as established by a social capital survey instrument baseline measure?*

4. *What are the implications of improved, behaviour-based measurements of social capital levels for urban and social planners?*

The unique data set examined in this project allows a direct comparison of movement patterns data and social capital data across a sample of 97 participants in a case study that enables at least provisional answers to these core research questions. The structure of the research from the philosophical framework to a reflection on how the insights gained are important for planning and policy development is provided.

1.5 Definitions of Social Capital

Social capital has a long conceptual history while the term itself, though occurring a century ago, began to grow notably in academic circles in the 1980s (Farr, 2004) and in popular culture since 2000 (Watts, 2004). Diverse thinkers and practitioners have been involved in social capital research and applications leading to dozens of definitions of social capital in the literature. There are earlier core lists of definitions that are useful (Adler & Kwon, 2002) but given ongoing development, incomplete. A selective review of social capital definitions from leading scholars in the history of social capital provides context for the definition I have developed for use in the current study and should be understood as representative rather than exhaustive.

A structural definition of social capital does not consider the phenomena such as trust or reciprocity but only includes the structure itself (Woolcock, 2001, p. 12) rather than the individuals within that structure (Lochner et al., 1999, p. 260). These structural and personal ties are better understood as complimentary and interaction (Belliveau et al., 1996, p. 1572). The ability of an individual to pursue interests by means of social ties may include shifting those relationships (Baker, 1990, p. 619), accessing commonly held resources (Finsveen & van Oorschot, 2008, p. 295; Woolcock, 1998, p. 198) and mobilizing them in a desired direction (Campbell, 2013, p. 80; N. Lin et al., 2001, p. 32). Mutual trust is often a pre-requisite for this kind of access (R. D. Putnam, 1995, p. 67; Rostila, 2011, p. 321) and may serve as a basis for a healthy society (Farr, 2004, p. 9; Uphoff & Wijayaratna, 2001) through collective action (Brehm & Rahn, 1997, p. 999). An empirical feature such as the size of a social network (Boxman et al., 1991, p. 52) can be expanded to include the assets that the actors in the network can mobilize (Nahapiet & Ghoshal, 1998, p. 243) including voluntary cooperation (Inglehart, 1997, p. 188; C. Y. Thomas, 1996, p. 11) often exercised in local settings (Jacobs, 1959, p. 180). These actions can have governance, sanctioning and conflict resolution dimensions (Ostrom, 2005, pp. 255–288).

Closely related to trust is the idea of reciprocity that reflects social obligations arising through interactions (Knoke, 1999, p. 18) and acquired obligations (N. Lin et al., 2001, p. 6). These obligations also imply economic aspects of resource deployment (Burt, 1997, p. 9), economic growth (Pennar, 1997, p. 154), and the influence of people on each other in pursuing goals that have economic implications (J. Coleman, 1994, p. 302; Alejandro Portes & Sensenbrenner, 1993, p. 1323). These economic dynamics may function as utility functions in formal economic evaluation (Schiff, 1992, p. 160).

These economic benefits extend to a range of collective resources needed for individuals to thrive (Woolcock & Narayan, 2000, p. 3) and secure benefits for themselves and others in their networks (Loury, 1992, p. 100; A. Portes, 1998, p. 6). This wider network includes groups pursuing common interests (Fukuyama, 1996, p. 10) through shared norms and values (Fukuyama, 1997) that range from personal to large scale societal belief systems (Fukuyama, 2002, p. 1). Institutionalized forms of collective activity are still dependent on trust and reciprocity as noted but are more fully formalized (Bourdieu & Wacquant, 1992, p. 119) and may include status-oriented institutional titles (Bourdieu, 2008, p. 243). Both the institutionalized structures and individuals within it combine to form social capital (Bourdieu, 2008, p. 248). Lack of agreement on a formal definition suggests both that social capital research may still be early in its development (T. Kuhn, 1962) and is a phenomena too complex to define in simple ways. Core elements across the range of definitions include trust and social networks (Richard M. Carpiano & Fitterer, 2014; Chow & Chan, 2008; Richey, 2007; Veenstra, 2002). It is important that the core elements of a social capital definition identify the core aspects of the phenomena (Dugundji et al., 2012; Kwan, 2007; Logan, 2012; van der Gaag, 2002). The working definition of social capital that I am using refers to: *the density and quality of human relational networks both formal and informal*. The density refers to the number and kind of relationships (how many close friends, family, acquaintances) and the quality refers to the levels of trust in those relational ties, including levels of trust in larger structures such as institutions.

Lower levels of trust lead to increased relational fragmentation, social isolation and hence decreased levels of social capital (Bethune, 2014). Social fragmentation can increase pressure on municipal resources through increased litigation, greater involvement by police or municipal enforcement in disputes, low voter turnout and volunteering, and low levels of informal interaction with consequent increased health service demands (Browning et al., 2004; McDoom, 2014). For example, research in Holland points to increased social capital as an effective means of lowering the costs of public health service while improving positive results for patients with Parkinson's disease (Tod et al., 2016).

1.6 Planning and Context for Social Capital

Cities have always been sites of negotiated differences – whether in the experience/conception tensions noted in the previous paragraph or in the more direct challenges of who will live/work/play where in any given community today. Specific human location and the perceptions of those humans about their social and spatial worlds remains a rich space for exploration. I make no pretense of fully examining the full range of those possibilities. On the contrary, I have chosen to focus on a very specific axis of interaction narrowed to a fine enough point such that preliminary conclusions might be drawn around the central and secondary hypotheses.

Community cohesion and belonging are critical for people living in cities today and requires that, in addition to our families and friends, acquaintances (people who know each other in more passing ways than would be the case with relatives or close friends), have access to and utilize settings where they can experience non-intensive social contact with people who are not like them - difference with cooperation. This negation takes place in community building encounters across a range of semi-public places like libraries and community centres (Peterson, 2016).

Difference occurs as well within what may appear to be homogenous social or cultural groups. Hyper-diversity is a phenomena where people who appear to have a great deal in common (e.g. an immigrant community in a given area of a city) sub-divide on the basis of very fine-grained preferences or differences (Peterson, 2016, p. 1069). Urban density provides more opportunity for a wide range of preferences to be met but can simultaneously pose a challenge for those seeking to build stronger community connections. Preference-seeking dynamic are operative not only within identified communities but across larger social environments such as neighbourhoods and can lead to various kinds of segregation even where dislike of difference is not present (Schelling, 2006). Both experience and research affirm that differences can be bridged but should not be assumed (Schurmann et al., 2002). Social capital research linked with locations supports a greater understanding of such dynamics.

Neighbourhoods are a city scale where human encounters across differences occur most often – we drive, walk, bike or run past other people at a neighbourhood level – and where we utilize semi-public places.1 Cities may be very large but human encounter can't occur at the scale of a whole city. Rather, people see and experience other people in stores, bars, commercial buildings, community spaces, places of worship, civic buildings and sidewalks (Hillier & Hanson, 1989; Piekut & Valentine, 2017).

[1] As noted earlier, 'neighbourhood' has a wide range of definitions and may include geographically complex spaces but here we may think of it as a space experienced by an individual across a full range of their activities on a day-to-day basis. The geographic scale of Census Tracts have been used in this study as a convenient proxy for such a scale.

This does not mean people will always interact in pro-social ways as noted above in the consideration of hyper-diversity. For example, social and even spatial patterns are being shaped by emerging cultural trends such as online and virtual exchanges that appear to amplify our self-directed preferences and perceptions of ourselves and others rather than diversify our social worlds (Amin, 2002). Amin emphasizes that most of this negotiation takes place in very local settings, whatever else may be going on at larger scales. We are, to some extent, forced to face the tangible, specific, spatial realities of other people in local, neighbourhood scale spaces.

The philosophical and theoretical examination of space has a long and rich history, including the work of people like Henri Lefebvre who propose that space has three important dimensions: perceived space related to the everyday interactions we experience and understand, conceived space that has stronger professional dimensions familiar to developers, planners and academics which fits the experiences of others (e.g. a community of people in a low income area) into a wider context and finally, an imaginative space where the mind of an individual locates their ideas, experiences and future possibilities into an imaginative whole – a third space (Lefebvre, 1992). Lefebvre is one example of an influential figure in planning theory that bridges the practical and the speculative encompassing a wide range of views about how the social as humans experiencing desire (Yamamoto, 2018) shapes and is shaped by much larger societal, cultural and structural dynamics (Pacheco, 2018). These realities – both empirical and conceived – imply deep social and spatial influences.

For as long as urban planning has been formalized, planners have had to work with the dynamics of social and spatial tensions generated by our human differences. Gill Valentine has argued that the 21st Century will be about living with differences (Valentine, 2008). Valentine points out that the context and effect of these differences is seen very differently by different urban theorists and practitioners who may advocate for various shades of cosmopolitan contact theory ranging from throwntogetherness (Bork-Hüffer & Yeoh, 2017; Doreen Massey, 1995) to common civility (Nigel Thrift, 2005) alongside many other gradients of thought.

Additional contemporary research on the sociality of urban spaces reflects increasing complexity in both our experiences of streets, shops, cultural venues and transportation paths and how those experiences are shared and interpreted locally and globally via digital avenues. Digital sources of information that are not local play a significant role in shaping the perceptions of individuals (Landy et al., 2018;

Navarrete & Castillo-Ortiz, 2020; Takahashi & Tandoc, 2016). The nature of these social network/trust bonds, ties, and external links are central to social capital research and evaluation. The simplifications pursued in my research are not intended to be conclusive by reduction but rather to open up the possibility that a specific, formalized examination of pattern-detection in social-spatial dynamics – reflected in the idea of social capital – would be valuable for urban planners if it is detectable.

As a result, the research design of this project does not include detailed examination of individual social or spatial habits. Those individual realities are vital for larger scale social and spatial patterns but since planning is undertaken at an aggregate level, the social-spatial interactions also need to be examined for insightful patterns at that scale. Individuals form multipoint centers of activity in their day-to-day lives that are still not fully understood from a planning and use perspective (Hasanzadeh et al., 2021) although we know that mobility is a critical factor in the changing ways we use cities and space today (Hasanzadeh et al., 2021, p. 2).

It is common enough to see reference to the term 'encounter' in both planning literature and planning practice documents but the meaning and use of that term remains unclear and in need of further definition (H. F. Wilson, 2017). By focusing on the specific dynamic of social capital quantity (as measured by the survey instrument based on the General Social Survey) and a range of spatial movement characteristics (captured by the empirical GPS tracks) at a neighbourhood scale, I have sought to enrich ongoing discussions of dynamics like 'encounter' without diverting down yet another difficult path.

Pursuit of a more detailed study of the scenes of encounter would not be antithetical to what I have examined here since relation (social network) occurs in space and place (geography) contribute powerfully to how we perceive and experience 'the other'(H. F. Wilson, 2017, p. 453). That such events take place in a neighbourhood and involve history, personality, social location, tourism, and a complex array of psycho-social scales and images is beyond dispute, ranging across concepts of borders or divisions (Rovisco, 2010) to long-established colonial settlement patterns (Leavelle, 2004). My assertion is that a specific aspect of this dynamic can be identified and explored formally.

Reviewing the range of literature related to these dynamics has convinced me that planning needs a means to take account of social and spatial patterns in a way that can be used in day-to-day planning decisions or evaluations. It is possible to examine a neighbourhood level effect without fully detailing all the dynamics that contribute to it. The use of mono-centric, bi-centric and polycentric activity centres across age and income differentiated groups suggests advances in insight can occur by aggregating individual patterns (Hasanzadeh et al., 2021, p. 8).

Significant attention has been given to studying racial tensions in the US at a neighbourhood scale and has led to a range of insights about how different groups experience each other and the communities they live in. Two people from different ethnic groups can have very different perspectives on the same neighbourhood (Walton, 2018). One telling dynamic is the degree of agency that an individual perceives they have over the events and decisions made in or about their community (Syme, 1991) and how that relates to mental health and community participation (Whitehead et al., 2016). Since social capital is comprised of a complex mix of social networks, perceptions of trust, variances of scale, and geographic location, issues such as racial, cultural, and socio-economic divides will only begin to be addressed and the current political tension eased in countries like the US when our understanding of local scale dynamics is advanced.

There is an important relationship between in-group and out-group trust and the social network data collected in the survey developed for this study. In-group and out-group trust is a well-developed body of work that examines the kinds of tensions outlined above. Trust is a central feature of social capital and is experienced perceptually by individuals. We know that such perceptions have a 'contagion' effect where like attracts and fosters like. For example, happiness / unhappiness not only affects the individual experiencing it but has been shown to influence others in the social network up to three degrees away (Fowler & Christakis, 2008). Happiness, like trust, is neither randomly or nor evenly spread but tends to cluster. These effects do not seem to percolate as strongly through work environments, but family, friends and spouses do correlate in terms of happiness, impacted more by bonding than bridging capital, it would seem. These network effects decay with geography and time – more of both erodes effects for both happiness and unhappiness (Fowler & Christakis, 2008).

Differentials along an ingroup and outgroup axis have been studied to determine how our personal experiences relate to our perceptions of the wider society of which we are a part (Delhey & Welzel, 2012). Delhay (et al) reports that ingroup and outgroup effects are not symmetrical in terms of influence. Ingroup has some impact on outgroup trust (ingroup being those close to us – specific trust and outgroup representing trust in institutions, strangers, or society at large – general trust). The non-symmetry of these dynamics, including the decay / influence noted in the previous paragraph, strongly suggest the complex nature of social capital. This complexity does not render the phenomena opaque as patterns and tendencies have clearly been shown.

One of these dynamics is the challenge represented by radical political groups that form within wider democratic societies. The tendency to abandon mainstream or accepted pathways for political action has lost ground to approaches that are much more radical in nature. It has been asserted that populism (whether on the political right or left) represents a failure of social integration (Gidron & Hall, 2020). Whatever your political preference, if you are on the margins, you are much more prone to radical positions. This seems to stem from feeling powerless to influence changes through 'normal' channels thereby leading to a perception that more radical pathways are the only viable options. When there are opportunities for involvement that matters meaningfully and concretely, radicalization tendencies are decreased (Gidron & Hall, 2020, pp. 1030–1031). For those who feel 'left behind,' mistrust and resentment lead to consideration of any path that is open, including a radical option.

From a social capital vantage point, perceptions of powerlessness would show up as low trust in institutions or societal structures and high trust in a much smaller personal network (or perceived as personal) where an ability to enact change is present alongside others who share that perception. It could be that such differentials, if easily measurable, are markers that signal a need to adjust policy or planning approach. In both cases, a social or collective dynamic is involved but the scale and scope of that collective space can be dramatically different. Marginal groups are known to have their own factions and in/out dynamics but can, in some cases, effect change as a very small minority – e.g., the relatively small number of people involved in the 9/11 incidents dramatically shifted national and international policy for billions of people.

While it is tempting to look at current or recent events to affirm this, the pattern has a long historical trend and has been tracked, for example, in Europe by various indices (Rohac et al., 2017). Prior histories of marginal group actors continue to influence contemporary actors through a struggle for power at political, media and institutional scales (Ten Brinke & Keltner, 2020). A wide range of strategies are used with a power differential (perceived and/or real) providing the conditions for changes of thought that can lead to concrete actions. These differentials can be identified across scales – rejection of a family group in favour of a new association or at national or international levels through various political groups (both formal and informal). Those who have power (high social capital, in our case) tend to see collaborative strategies as the most effective and legitimate while those without power (low social capital) tend toward strategies that disrupt and seize power, including through violent action (Ten Brinke & Keltner, 2020, pp. 8–10).

From a planning and policy vantage point, it is imperative that people who work with institutional framework (as do most professional planners and scholars) recognize that perceptions of power and trust, captured by measures of social capital, relate directly to the efficacy of strategies – current and proposed. By this line of reasoning, for example, it would be expected that neighbourhoods (Census Tracts) with low social capital (particularly on the institutional side) will not perceive city-run community consultations as trustworthy, de facto. The many aspects of social capital relevance for planners noted in the preceding suggestion suggest the importance of research investments that seek to understand it more fully. To support that research insight, I will devote the following section to a review of the history of social capital.

1.7 History of Social Capital[2]

The phrase "social capital" can be traced back to the mid-19th century (Tocqueville, 2001) but the concept of people acting together to achieve what they could not do alone is ancient (Aristotle, 2012). One of the drivers of interest in social capital is a growing awareness of how much the nature and intensity of our human relationships shapes our lives at all scales from the interpersonal to the global (Franke, 2005). This interest has included the Federal Government in Canada where deployment of public funds and programs toward solving social problems constitutes significant resource and infrastructure commitments. Academic, policy, and program specific reviews reflect the challenges and opportunities of social capital research.

Prior to 1993, it is possible to trace a paper-to-paper linkage of key works on social capital and to consider their various themes and ideas. This identified range ((Ostrom & Ahn, 2003) represents a 153

[2] A more complete version of this section was published as: *Is Social Capital Measurement Still Relevant? A Roundtable Policy and Social Impact Assessment White Paper* by Milton J. Friesen. 2014.

year span of time from the early use of social capital found in *Democracy in America* (Tocqueville, 2001) through to *Making Democracy Work* (R. D. Putnam et al., 1994)(Table 1-1).

Year	Name	Title
1840	de Tocqueville	Democracy in America
1916	L. J. Hanifan	"Social Capital – Its Development and Use"
1961	Jane Jacobs	The Death and Life of Great American Cities
1961	Shultz	"Investment in Human Capital"
1962, 1964	Becker	"Investment in Human Capital: A Theoretical Analysis" and *Human Capital*
1973	Granovetter	"The Strength of Weak Ties"
1977	Loury	"A Dynamic Theory of Racial Income Differences"
1986	Bourdieu	"Forms of Capital"
1988	James Coleman	"Social Capital in the Creation of Human Capital"
1992	Ronald Burt	Structural Holes: The Social Structure of Competition
1992	Ostrom	Crafting Institutions for Self-Governing Irrigation Systems
1993	Putnam	Making Democracy Work

Table 1-1 Selected social capital scholarship highlights spanning 153 years (book titles in italics).

The following is not a comprehensive review of all uses or related uses of social capital but is instead a review of significant growth in social capital interest. Durkheim (1888) is an example of a late-nineteenth century use of social capital where a discussion of the conditions of suicide were under consideration (X. Chen et al., 2009). In the second half of the twentieth century there is an increase in the number of academic works on social capital but it was a pivotal paper on human and social capital in the late 1980s that accelerated scholarly activity (J. S. Coleman, 1988). One of the contributing factors in Coleman's influence was the way in which he integrated insights about both individual human choice and agency (including economic ideas about rational choice) and the role of larger social structures (groups, organizations, institutions like education, and so on) in both constraining and enabling human action.

A few observations about this traced pathway will provide illumination of the point. L. J. Hanifan (Hanifan, 1916) published a paper describing how a local school could function as a community centre and in so doing, contribute to the overall social resources, the *social capital*, of the community in which it resided. The idea that a common pool resource in the form of a collective social good could be generated through the functions of a community institution was not new but describing it as a form of capital, as social capital, was distinctive. However, prior to 1993, Hanifan was simply not cited by scholars and is a clear example of a "sleeping beauty" paper that suddenly wakes up (Ke et al., 2015; Raan, 2004). Despite his early adoption and description of the term social capital, Hanifan was largely ignored in the academic literature until an increase begins with 22 citations in 2005, 57 citations by 2010 and 103 by 2014 (cf. Web of Science review – details noted below).

This obscure scholarly paper re-discovered by contemporary researchers is indicative of growing scholarly interest in social capital research. There are other facets to be considered in this early period, including the work of urbanists and planners. Jane Jacobs was an important social capital contributor despite being better known for her contributions to urbanism (Ostrom & Ahn, 2003). Jacobs was not a social capital scholar but observed the effects of collective action (or lack thereof) in urban settings and remains one of the most-read sources for planners and urbanists (Filion et al., 2007) along with economists and political scientists (Sancton, 2000). She is, however, scarcely represented in the WoS™ database as she was not a formal scholar who published in the academic literature. This must be countenanced by her extraordinarily wide influence in planning and urban development, including the social capital facets of city building and design in her work, particularly the less visible social aspects of vital communities. This is a reminder that not all contributions to social capital scholarship are included in a database. A more recent indicator of social capital ascendency is Schultz (1961) prominent paper on human capital which was published in the prominent journal *American Economic Review* and has a significant number of citation records (864).

After 1993, a surge of interest begins to fuel academic momentum at a scale that makes individual paper listing impractical. The building volume of scholarly work on and related to social capital forms a vast stream where identification of themes, subject areas, and categories becomes essential as a strategy for reviewing the literature. Based on the sheer numbers, the top 20 themes in WoS™ represent a significant range of academic tributaries that can be examined in their own right to identify the primary avenues of research and writing on social capital. I will now examine that scholarly context more fully in order to support one of the central tenants of this project, that social capital is a significant phenomenon across a range of planning-related fields and warrants intensive research.

1.7.1 Current Academic Research on Social Capital[3]

Current academic interest in social capital is robust and extensive, ranging across dozens of disciplines from economics to geography to medicine. This section reviews that the scope of that landscape. Specific aspects relevant to each of the following chapters will be identified as they support the examination of the hypotheses.

The Web of Science Core Collection™ (WoS™) contains a constantly expanding repository of eighty-one million academic papers, books, conference proceedings and citations across the social and natural sciences. Within that repository, social capital is well-represented with more than 21,198 records as searched in July 2019 where the "social capital" search phrase is in the title, abstract or key words.[4]

1.7.2 Research by Discipline

The wider list of WoS™ Research Areas reflects significant interest from business/economics, sociology, psychology and a range of social science and behavioural science disciplines. These categorical patterns are only part of the wider landscape pattern in social capital scholarship.

---

[3] This is based on an earlier published version of the meta-review that used 2014 data and can be found here: Friesen, M. J. (2018). Navigating Social Capital Scholarship: A Meta-Review of 9164 Web of Science Sources on Social Capital. *Cardus Research Report*. Available: https://www.cardus.ca/research/social-cities/reports/navigating-social-capital-scholarship/

[4] July 4, 2019 – Web of Science Core Collection "social capital" as search parameter.

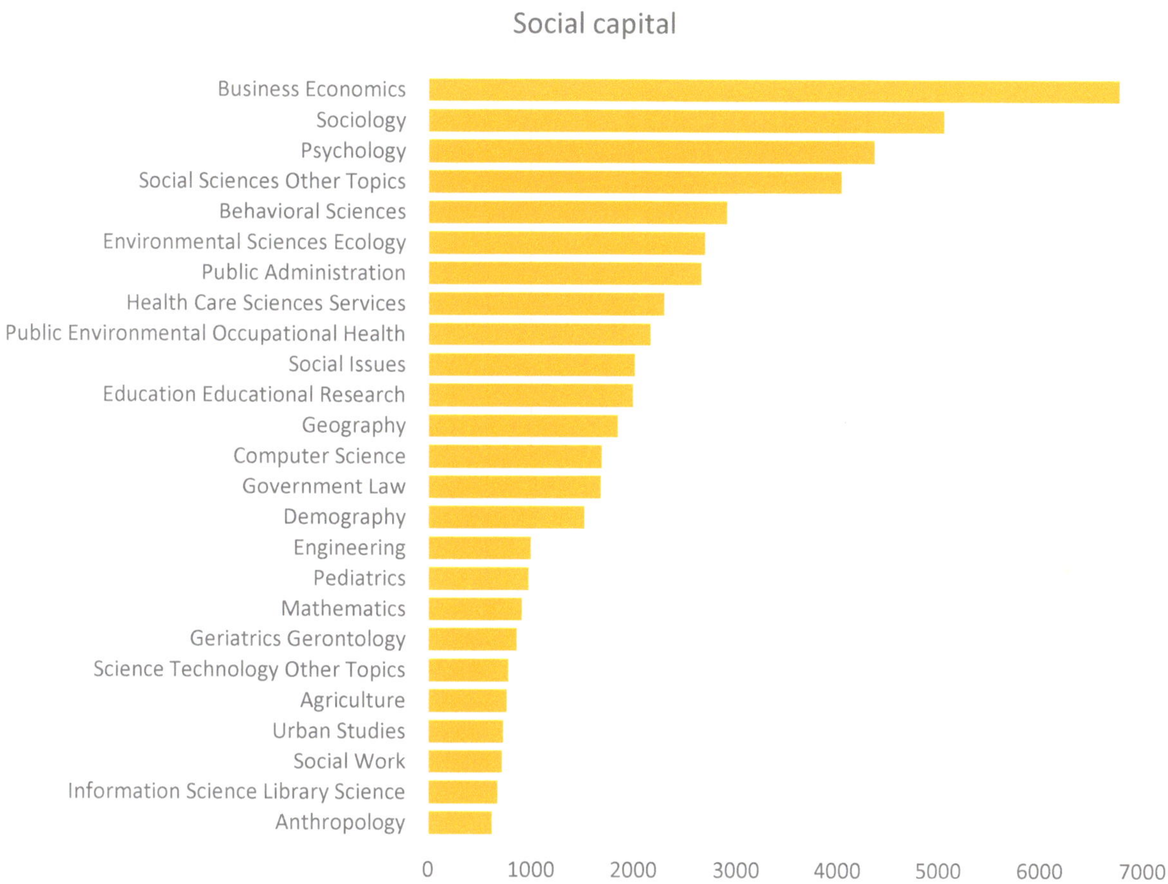

Figure 1-1 - WoS™ Research Categories by number of records.

Citation links of WoS™ Categories represented by *social capital* reflect direct scholarly interaction including significant inter-connectivity across disciplines (Figure *1-2*). Colour and node location within the network indicates the number of papers within various disciplines and fields of study. The network shape is affected by the number of citation links a given node has. The size of the circles (nodes) indicates the relative number of records, and the connections are the number of citations.

This citation network shows that social capital scholarship is complex, multi-disciplinary, significant use of social capital can be clearly seen by colourizing the WoS™ Categories by domains (nodes) and the citation links (edges) that connect them. Zooming into the network map demonstrates that many planning scholarship domains are included in this research network (Figure *1-2*).

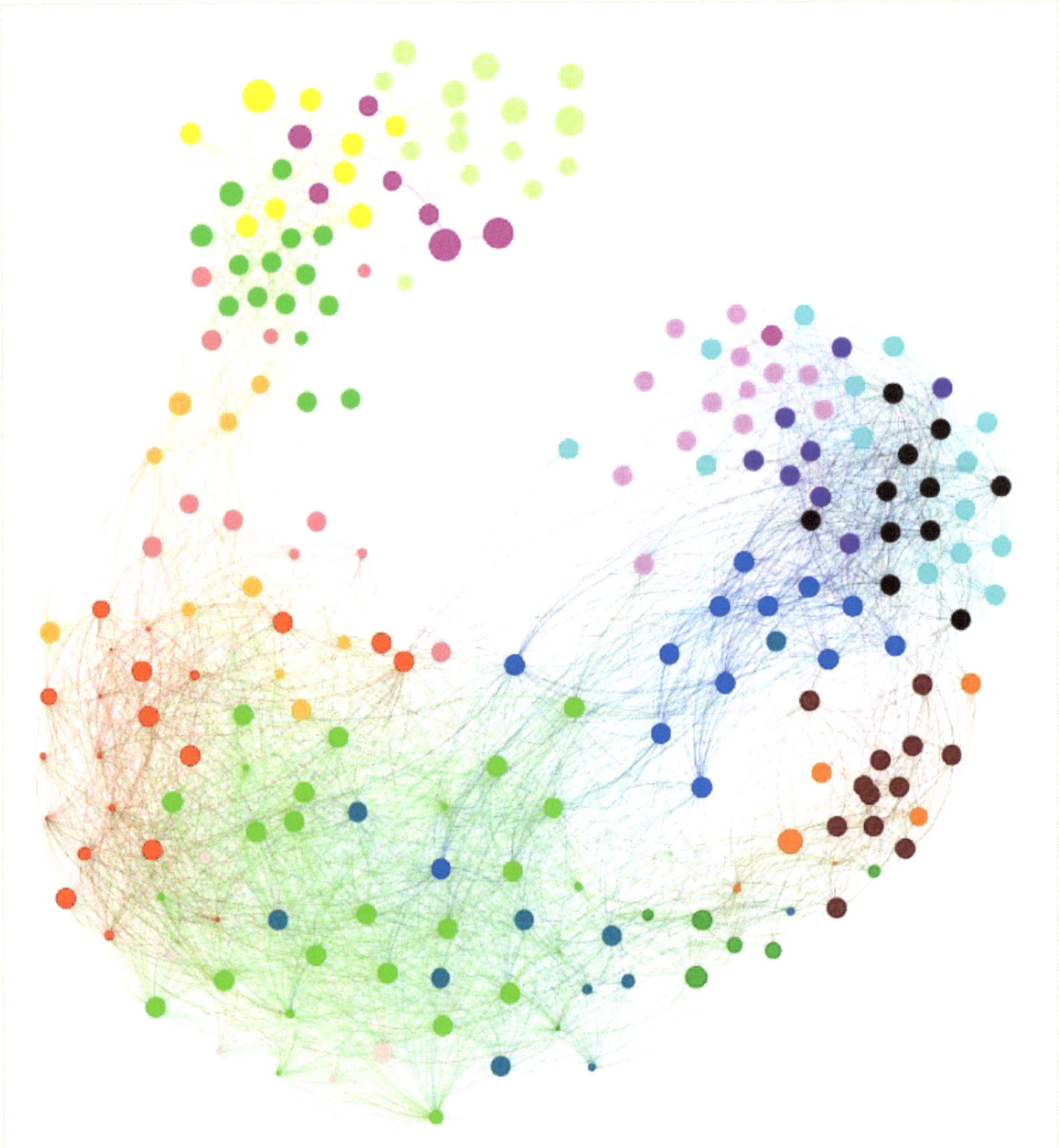

Figure 1-2 - WoS™ social capital coloured citation graph by categories. Red is medicine, orange is Psychology, light green (lower section) is biology, brown is Earth, Environmental, Engineering, yellow at top is sociology, lightest green (top) is planning and public administration, etc. Citation data was extracted from WoS™, converted to .gexf format using an Excel macro and then imported into Gephi for visualization and analysis. Square is area highlighted in the following (Figure 1-3).

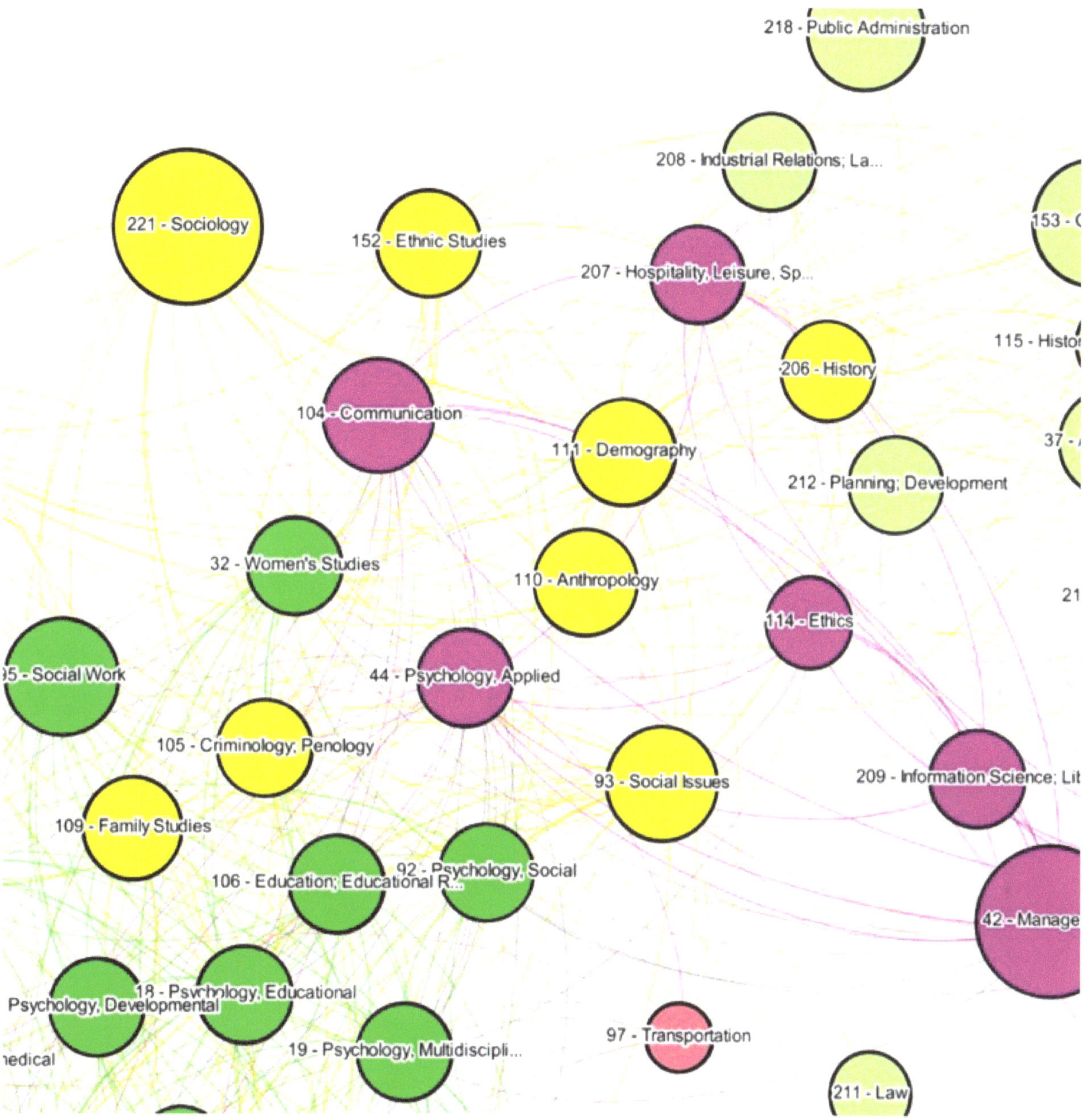

Figure 1-3 - Magnification of a selection of the WoS™ social capital coloured citation graph by categories.

Visualizing the top 500 most cited sources by author with the CitNetExplorer software provides a view of the works cited (links that go up) and works that cite the given source (links that go down). This citation network traces paths of individual scholarly influence. Coleman (1988) is near the centre but still has many more links down, reflecting that Coleman is highly cited but does not cite older sources at the same rate. CitNetExplorer uses colouring to indicate clusters based on bibliographic coupling – the more they cite sources in common, the more centralized they become (Figure 1-4). In network theory this modularity is reflective of groups within the data where more ties means nodes closer together (Newman, 2009).

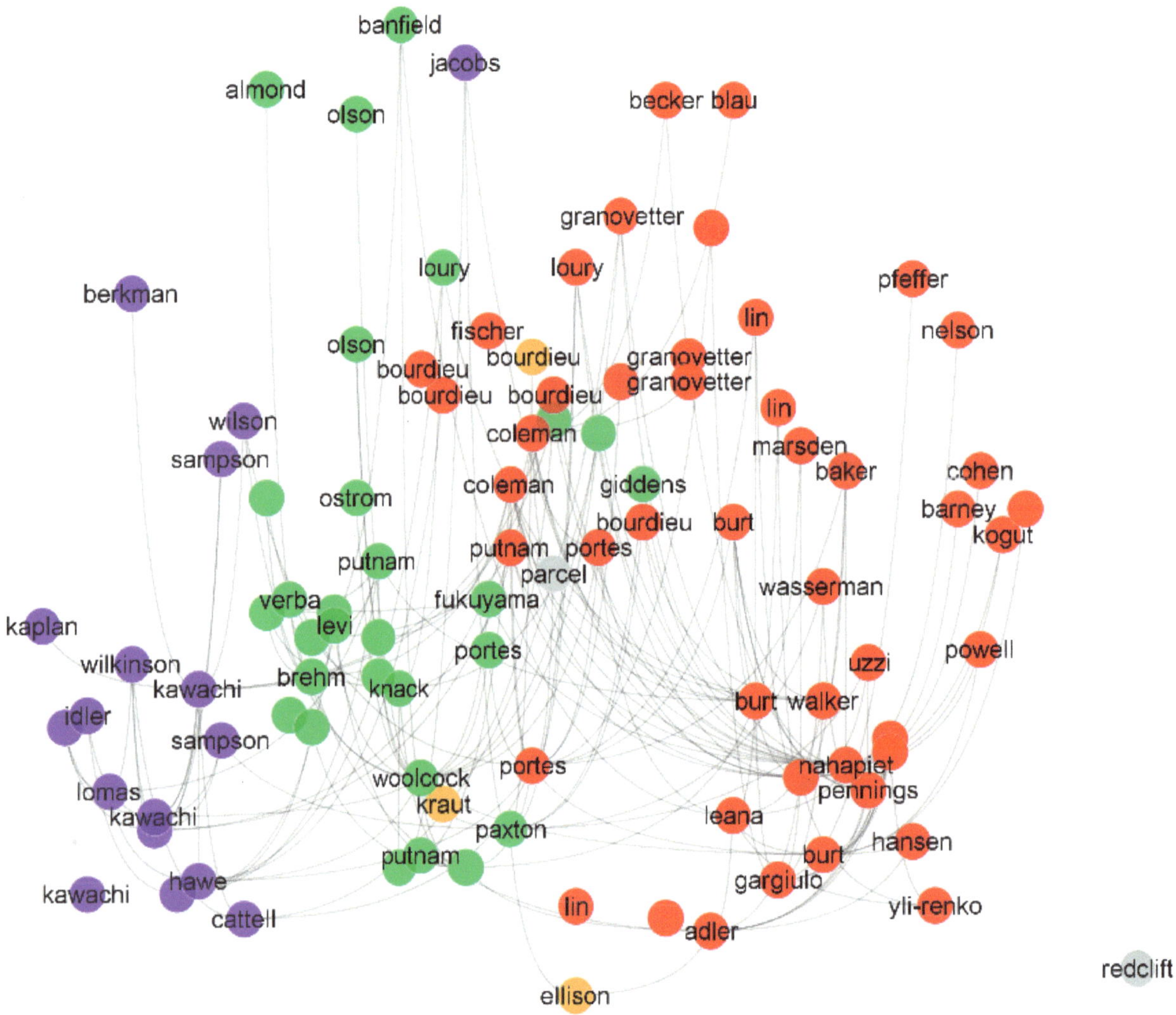

Figure 1-4 - 100 citations from top 500 most cited (limited to 100 nodes for clarity). Groups formed by citation closeness. In this case, the oldest paper is at the top Banfield (1958) and the newest is at the bottom Ellison (2007). Data downloaded from Web of Science August 29, 2019.

1.7.3 Research Across Time

Prior to 1991 the number of publications demarcated by *social capital* was less than 2. Steady increase in interest continues to the present with a doubling of records from 1997-1998, another doubling by 2002, and again by 2007 (Figure *1-5*).

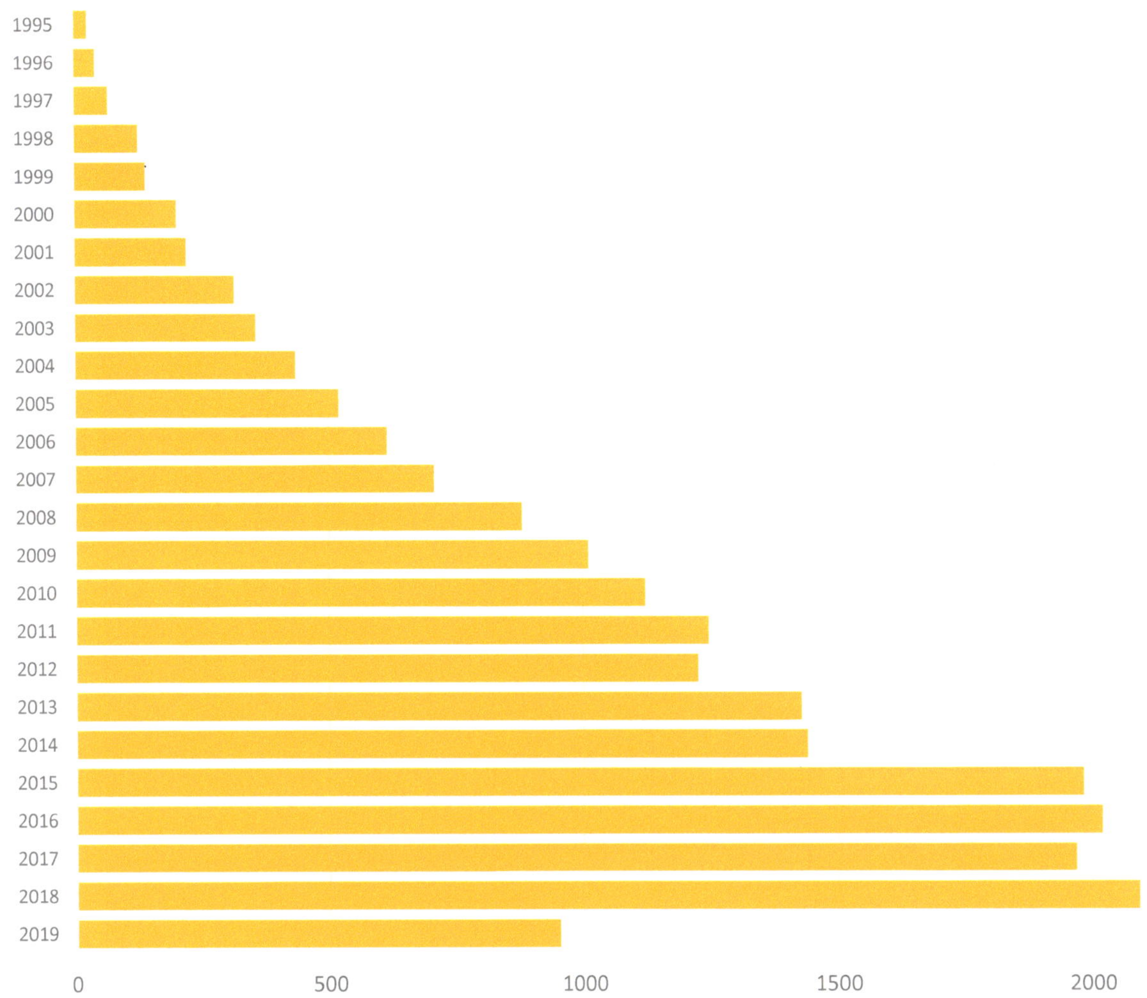

Figure 1-5 - Publication of social capital papers by year.

1.7.4 Research of Abstract Themes

Within the full collection of more than 87,000,000 records that comprise WoS™ (July 2019) there are more than 21,198 records that are returned by a *social capital* search. Abstracts were downloaded and combined into a master text file that comprised 3,575,756 words (24,255,531 characters). In what follows, I use text mining and citation techniques to thematically review the breadth and inner relationships in both the citation and abstract text files. Analytic descriptions of the finer grained approaches within this wide literature is undertaken by examining an important sub-category of social capital that is directly relevant to the central hypothesis that individual social capital levels increase with increasing movement in a city.

The relative sizes and key counts for the textual database are summarized in (Table 1-2).

Data File	Number of Abstracts	Unique Words	Unique Words Post-Processing	Words Retained for Analysis	Frequency of Words for Graphing
“social capital”	21,198	32865	16384	50%	3000
“spatial”	187	4738	3146	66%	60

Table 1-2 Summary of WoS™ social capital textual data with “spatial” subset.

The nearly 21,198 abstracts in the core data set comprises, 3,572,818 total words and 64,118 unique words. Manual review and analysis of this volume of text content is impractical. In order to maximize information within the abstracts, text mining tools were employed to determine the statistical shape of the body of the combined abstracts.[5] Themes relevant to urban planning scholarship are evident and include *health*, *community*, *trust*, *development* and *networks* (Figure 1-6).

[5] For reproducibility, R code and text data file used are available upon request.

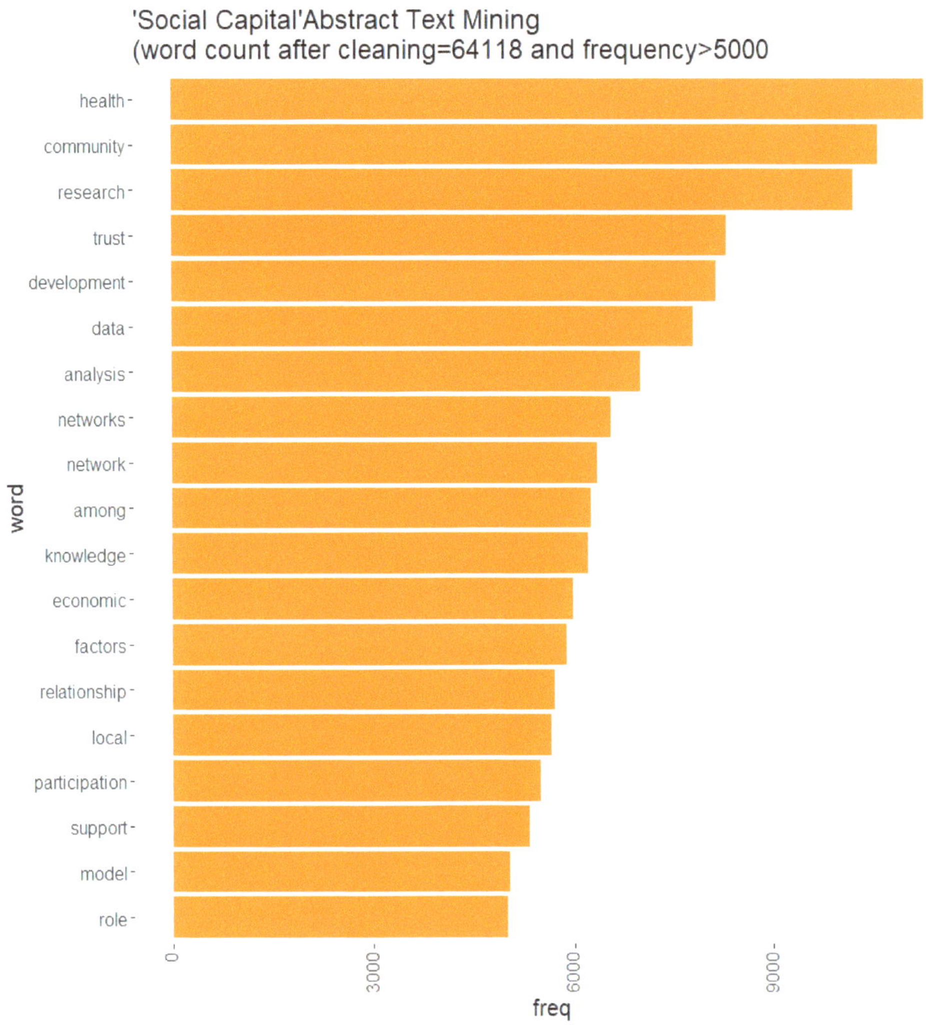

Figure 1-6 - Web of Science social capital abstracts with words occurring more than 5000 times arranged by frequency. [Raw text was processed in the following way: punctuation removed, numbers removed, converted to lower case, stopwords removed using "english" stopwords from R (n=174), removed definition words ("social", "capital") whitespace removed, and frequency matrix generated for analysis].

Spatial analysis related to social capital is central to my examination. When the abstract text analysis is refined by adding "spatial" to the search terms in WoS, similar themes emerge with some variation in order: *development*, *economic*, *local* and *urban* are more prominent and *health* is less frequent (Figure *1-7*).

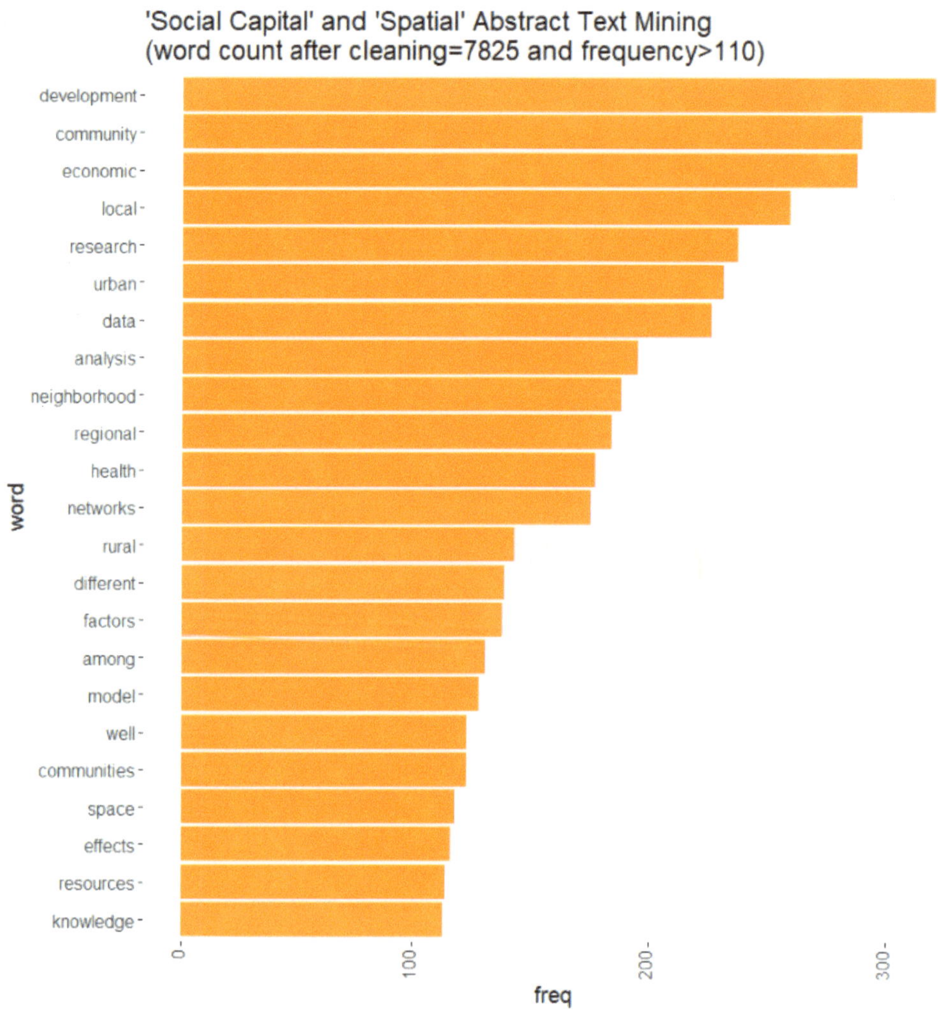

Figure 1-7 - Word frequency for abstracts that contain both "social capital" and "spatial" as search terms for WoS.

1.7.5 Significance of Social Capital Research Overview

The current state of social capital scholarship as reflected in the vast WoS™ Core Collection™ leads to five observations relevant to the current study:

First, *sociology* and *business/economics* are important disciplines within which social capital research is active and growing. Both are important fields of study for planning scholars and practitioners. Sociology has a long history of the influence of spatial use and movement on human behaviour. *Health* is also a very significant field as well, particularly the role that social structures and social context play in human well-being which is also highly relevant in a planning setting generally as public good (or common good) tenants are expected of those engage in professional planning duties (American Planning Association, 2012; Canadian Institute of Planners, 2012).

Second, the field of social capital is made up of both a small but prolific group of authors (primarily in the health field) and a significant number of much less prolific authors. This represents a challenge that scholars reviewing the field need to be aware of. Reviews of only the most cited or most prolific authors leaves significant scholarly work under-used in the long tail of small but numerous contributors. Cross-disciplinary approaches, by implication, are essential for social capital scholarship and reviews of literature cannot be narrowly focused without losing important contextual contributions.

Third, the USA is at the forefront of English language social capital research with Canada and the UK supporting that trend. China, Germany, and the Netherlands are also notably active along with a significant number of countries making more modest contributions. English language analysis is a limit as work in French (particularly coming out of Quebec but also from France) and other communities would increase the range and potential insight of existing scholarship.

Fourth, published academic articles are *the* format for exchange. Though there are several books and book reviews, they are marginal by comparison with journal articles. This may indicate that the field is still emerging. It also suggests that translation of academically driven publishing in social capital to formats more useable to policy makers, government bureaucrats, and other decision makers is needed. It is probable that key methodological papers will emerge as citation and influence leaders in a way like James Coleman's seminal role in the late 1980s. Looking for new measures, proxy measures, and approaches to data analysis will represents a significant amount of potential work for a global community of social capital scholars.

Finally, building on point 4 above, substantial social capital growth has occurred in the last decade in the academic field but has had mixed uptake at policy levels depending on country. In Canada, for example, policy interest was very high up to 2004 (Franke, 2005) but declined significantly after that despite scholarly growth. It is likely that academic research will continue and perhaps grow even more diversified as population growth, societal complexity, and institutional change drives our need to understand how our collective social resources grow, decline, or change their character over time. Significant effort will need to be invested to make that work available to those outside of academia, including planners working in both private firms and city administrations. Given the complexity of social capital, it is important to position the research of this study in an adequately grounded framework.

This review of the scholarly landscape of social capital provides a context for the output side of research. I will now focus on another critical dimension of social capital – the conceptual structure that informs both the methodology and application of the project.

1.8 Social Capital Research from Philosophy to Application[6]

The complex nature of social capital – empirically, historically, spatially, and socially – requires a clear philosophical articulation that connects theoretical deliberations with a specific research methodology including the collection and analysis of social and spatial data.

My research structure is designed as an armature (Hu, 2014; Kojima, 2014) structure that moves with sufficient flexibility to allow for novelty and discovery on the way (Alexander, 2003a; Westley et al., 2006; Wheatley, 2006). Movement from left to right (Figure 1-8) scales from meta-philosophical considerations to particular phenomena that can be investigated empirically. This logical structure supports exploration of a difficult concept like social capital and spatial interactions without being falsely reductive (Abbott, 2009; T. Kuhn, 1970).

CRITICAL REALISM >>> COMPLEXITY SCIENCE >>> NETWORKS >>> SOCIAL CAPITAL >>> SPATIAL DATA

Figure 1-8 - Progressive framework for social capital from philosophy to the application.

Critical realism provides a philosophical framework to advance social capital and spatial research and is oriented around four key concepts: *regularity in scientific knowledge*, *historical knowledge*, *human agency*, and *emergence*.

For knowledge to be generated, shared, and advanced, some degree of *regularity in scientific knowledge* is required. Although social capital may be a difficult phenomenon to explain, it is not

[6] The published version of this section: Friesen, M. (2019). Planning for Social Environments: Social Capital in the Context of Critical Realism and the Dynamics of Complex Systems. *Policy and Complex Systems*, 5(2), 221–242. https://doi.org/doi: 10.18278/jpcs.5.2.8

invulnerable to scientific advancement (Mir & Watson, 2001, p. 1170). Our knowledge, while always partial, can advance with new discoveries. Significant causal complexity and the deep interrelatedness of phenomena suggests that there are *in-principle* limits that will not be remedied by better research or more sophisticated computation and the non-linear nature of interacting dynamics results in a temporal reality that is never free of contingencies (Steinmetz, 1998; Taleb, 2010). This is something which planning and policy have increasingly recognized as a permanent feature of the environments they seek to organize (Allmendinger, 2009, pp. 18–23).

Historical knowledge (including social patterns that we can perceive through statistical data, direct experience and cultural knowledge) has always been used to guide decisions about the future. The data collected for this study is not live but was generated by participants in the past and may contain insights for both the present and future. These participants exercised *human agency* – the ability to act independently of external constraints. That independent action is limited by human capabilities at an individual level and is also constrained by larger order structures like laws and cultural norms. While we may choose to eat whatever we want, our location, resources, biological limits, and many other factors narrow what appears to be an open choice down to a narrow range from among a much larger possible field. Planning is premised on both individual human ability and built-in limits.

One of the important results of the tensions of agency and determinism is that law-like functions at one level can give rise to novel phenomena at higher orders; this is known as *emergence*. Emergence in this research context is "the relationship between two [structural levels of reality] such that one arises diachronically (or perhaps synchronically) out of the other but is capable of reacting back on the lower level and is causally irreducible to it" (Steinmetz, 1998, p. 173). The implication supports the limits on agency noted above in that both smaller and larger structures involved in social capital and spatial dynamics interact and that research design must give consideration to both individual and larger order structures. Emergence complicates our efforts to sort out direct causal arrows (Richardson et al., 2006). This implies that definitive or complete answers to social capital and spatial interactions are not likely. The relations between parts, elements, people, natural entities, and everything else around us is deeply complex at a causal level. At the meso level of social structural analyses (such as concerns investigations of social capital at neighbourhood levels), these patterns and regularities (law-like functions) may exhibit wide behavioural variations in reality and thus represent persistent but at least partially solvable challenges for research.

Complexity science provides disciplinary rigor for factoring the above noted dynamics into research design and is well suited to causally difficult phenomena. Social systems and structures arise from and shape human interactions at all scales including meso-level neighbourhood contexts in urban settings. Complexity science includes the study of non-linear, dynamic systems and processes where the number of variables and their nature preclude control through reductive analysis.

When Jane Jacobs wrote *The Death and Life of Great American Cities* in the late 1950s, complexity science as we know it today was still in its infancy. Jacobs identified an important 1958 review essay on scientific progress and complexity by Warren Weaver that signaled growing awareness of complex systems consequences generally and what such thinking might mean in the context of cities and their functions (Jacobs, 1959, pp. 428–448). Lindblom, writing a year or so ahead of Jacobs, reached similar conclusions about the limits of administrative planning. In principle, planning into the future as a way of fully controlling outcomes was simply not possible and could be detrimental where such plans failed to adapt to actual conditions over time (Lindblom, 1959). Cities are not machines, Jacobs argued, because they are characterized by non-linear processes implying irreducibility. Like natural systems of energy and motion, social structures and human systems seem to hold order and chaos in tension reflecting both predictability and novelty (Byrne, 1998; Lewin, 1999; Simon, 1962; Waldrop, 1993). One of the central features of complexity science focuses on how the discrete elements of a system relate to the whole (Humphreys, 2016).

Complex systems are not linear (Feigenbaum, 1983; Holland, 2014; Richardson, 2010) - very slight variations of initial conditions can yield remarkably different outcomes (Lorenz, 1963). This requires careful consideration wherever statistical analyses assume normal distributions are applied to complex, non-linear dynamics such as social processes. Complex systems are very interconnected and evolve through information feeding back into the processes that comprise the system, changing it as the information feeding back changes. This is a common feature of social systems at all scales (Feigenbaum, 1983). In social systems, such as those that constitute the conditions for social capital formation and function, people can and do change and influence it from within. If we are going to increase our understanding of and possible agency within social systems, we will need to account for complex systems dynamics in research design, models, and interpretive frameworks (Byrne, 1998). Planning is just such a form of intervention.

The step from the foregoing considerations to practical application involves the introduction of network science. Networks provide the linkage between critical realism, complex systems, and the empirical aspects of social capital. Analysis of complex phenomena within the scope of this study makes use of network analysis as a core, structural means of engaging with the complex social phenomena being studied. Core network science principles are used to support a framework for thinking about the central hypothesis and establishing a means of answering the resulting questions. Network structures, including social and spatial networks, are amenable to mathematical analyses across many dimensions making it possible to quantify social and spatial features which is useful for comparative study. In planning environments, transportation has made significant use of network analysis to understand the dynamics of how people and vehicles move on roads, sidewalks and other spatial structures and the role these structures play in large scale dynamics (Little, 2002; Omrani, 2015; Dominik Papinski et al., 2009). Two primary applications of network science are used here: respondent driven descriptions of their social networks (close friends, acquaintances, and relatives) and mathematical descriptions of their travel patterns analyzed with ArcGIS.

Social capital has no singular mechanism and no clear boundary between structure and dynamics. Interacting mechanisms, agents, and other variables are at best partly understood (Janssen et al., 2010; Ostrom, 1986, p. 16).

> It is not necessary to understand all of the underlying mechanisms of social capital in order to make provisional assertions about meaningful knowledge of social resources held among a number of actors – we can gain insight about social dynamics without proof of a causal link (Cushing & McMullin, 1989) between a given variable and social capital.

As noted earlier, there are tens of thousands of "social capital" papers and the subset of "spatial" is also significant. After refinement there are more than 414 relevant papers across urban studies (70), geography (178), public administration (95), and sociology (102) that combine social capital and spatial dynamics. Each of the identified fields in these papers are strongly related to planning. The text analysis revealed that the 7825 unique words (89,828 total words) found in the abstracts of these subset of papers reflected *development*, *community*, *economic*, *local*, *urban* and *neighborhood* themes among the most frequent occurrences – all themes that concern urban planning.

More specifically, recent papers within this group include studies of spatial patterning at neighbourhood levels (C.-Y. Lin et al., 2019), linkages between distance from public space and social capital resources (Foster et al., 2019), along with studies that examine spatial correlations between social capital and poverty (Harrison et al., 2019). Some of the most highly cited papers reflect that social connectivity and ecological preservation are important for resource management (Brondizio et al., 2009), that human health is shaped by urban design (Jackson, 2003), and that our lives at a neighbourhood level are better where social capital is higher (R. M. Carpiano, 2006). Public space, poverty, ecological preservation and quality of life are all within the scope of core planning commitments. These samples of scholarly work are a few among the many.

1.9 Methodology for Social Capital Research

The history, definitions and scholarly content of social capital examined above clearly indicate that social capital is an intricate and difficult phenomenon. The design of this study has considered those dynamics, particularly as they apply to the intricate social capital dynamics reflected in human interactions at neighbourhood scales. A research methodology sufficient for this complex phenomenon must balance specificity to advance research insight and flexibility to respond to what may be learned on the way. The research design offers supervised, unsupervised, linear, non-linear, and exploratory techniques. Each of these approaches are detailed in the following chapters as applied to a data set collected for this research design.

The data collected follows a traditional, disciplined sampling and survey approach. This structured approach supports the central and secondary hypothesis concerning the linkage between social capital and spatial patterns. While such approaches may suffice to enable slow, steady methodological progress, new forms of data that reflect phenomena such as trust, and social networks arises from the nearly ubiquitous flow of social data. This flow is generated by mobile devices, the Internet of Things, sensors, CCTV capture, transactional data, apps that integrate geolocation and human activity or exchange with strong social, spatial and technological interactions (Hampton et al., 2010; Keuschnigg et al., 2018). Some scholars argue that we are moving toward an N = all context for social and planning research (Mayer-Schonberger & Cukier, 2013). The pervasively social nature of this data reflects deep connections to the space of the city (location), the individual (as an individual, not an average in a group), and a deep shift from that individual relating to the group or public toward an individual to individual relational network that generates significant, ongoing, and accessible data (Carta, 2020).

For planners these data sources reflect a rich source of insight – first for data of this type that is publicly available through web scrapping, artificial intelligence tools, and administrative data collected from citizens (Hildebrandt, 2013); secondly through partnerships with for profit firms where data is made available in anonymized form (Bailey et al., 2019). Availability of social and spatial data at the scale we need (e.g. neighbourhood) but for entire cities – the idea of mass granularity – can still be challenging to source. The data exist but may not be in the right form, may exist behind commercial paywalls or legal boundaries, and may be subject to increasingly stringent privacy laws (Derman, 2011). The data may be plentiful but requires significant cleaning or it may be low quality (i.e. the data you want is a small part of a much bigger, messier data set) or it may not be the right data at all even if there is a lot of it (Wolfe, 2013). Finally, even if data is collected privately and under agreement, sharing results of research from that data may be limited by those data agreements – i.e., researcher may gain insight, but specific insights or representations of the data may be constrained given the inherently personal nature of most social data. Calls for examination of who benefits from the insight gained through the data appear to be more common. What are the motives of those collecting the data? How will it be used? How might the data increase the power of those who are already powerful (or degrade further those being exploited)?

If a state (city planners) can see us in very close detail, does that mean that they should? Researchers may be extensions of these systems and structures even if they are not conscious of how their actions deepen inequalities of power, access to knowledge, and the way in which the data may be used beyond the research framework, e.g., results may inform policy and political action that have nothing to do with the researcher. It is clear that our relationship with data, places and social structures is complex, evolving and will have a wide range of unintended consequences despite careful research (Beer & Burrows, 2013).

At their best, social and spatial data can help decision makers identify issues, pick up on public opinion at granular levels, development social return on investment measures and develop tools that close the gap between resources and people who need them. At their worst, these same dynamics are used to entrench power and further defraud the exploited.

GPS data provides the concrete linkage between human behaviour and the survey data (social capital) which can be analyzed using network science. This is consistent with our understanding of complex systems in the context of a critical realist philosophical framework. Recording the movement patterns of humans in their home territories provides an opportunity to examine how individual agency interacts with constraining structures related to geography, income, education, institutional experience, and a range of other perceptions. The following research reflects data about movement patterns including calculation of the distinct features of each participant. In the chapters that follow, survey data and GPS data are integrated and synthesized. This reflects both the structured (Borgman, 2015; Grimmer, 2015) and unstructured (Anthopoulus, 2015; Weinberger, 2012) data found in traditional surveys and the new data flows supported by smart city infrastructures.

1.9.1 Social Capital Measurement Frameworks

Ideally, policy relies on evidence and research support for decision making. In the case of social capital, it seems that policy appropriation of scholarly insight has lagged behind (Johnston, 2010; Ostrom & Ahn, 2003) but smart city, environmental and human centered design scholarship may be slowly changing that trend (Birk, 2016; Hollands, 2008; Lim et al., 2019; Reese & Rosenfeld, 2012; J. Thompson et al., 2017). Policy applications related to social capital have become more diffuse and now include themes such as social inclusion, civic engagement, and social resilience. Two central interests in social capital research are specifying core characteristics of the phenomena and identifying the most effective way to measure those features. The dozens of different approaches to social capital measurement indicate a lack of formal consensus on the most effective approach (as noted in the diffusion of definitions). In Chapter two, a range of measurement frameworks are reviewed based on scale of measurement, degree of acceptance, difficulty of application, and alignment with the social network and trust elements of social capital as outlined in the definition.

From among these options, the General Social Survey (GSS) was selected due to relevant social capital themes in the various cycles, its wide acceptance internationally, its potential as a source of formally evaluated and field-tested questions, and for the possibility of comparing the data from the present research with relevant data both nationally (Canada) and internationally through the International Social Survey Programme (ISSP). The rationale for using the GSS and the process by which relevant cycles were used as a source of questions for developing a new Social Capital General Social Survey instrument are fully explained in Chapter 2.

1.9.2 Planning and the Measurement of Social Capital

Planning has a stated aim of contributing positively to the public good (American Planning Association, 2012; Canadian Institute of Planners, 2012). Even if our understanding of social capital is incomplete, investing in understanding it better is important for urban planners who are responsible for designing and developing space that improves quality of life. Scientific work has always proceeded from partial understanding, competing theories, and unclear boundaries (Brody, 1970; Chomsky, 1996; Serres, 1995) so planners need not wait for complete explanations before applying research insight (some have argued that is the de facto mode for planning (Lindblom, 1959)).

Increasing our understanding of the social infrastructure of urban communities at the scale of neighbourhoods is particularly important for planners and policy makers. This local scale is related to a significant number of phenomena that contribute to well-being such as mental health, longevity, educational achievement, economic performance and belonging (Bethune, 2014; Dunkelman, 2014; Pinker, 2014). As an example, a social capital roundtable of experts was held in Hamilton, ON to examine the history, use and future evolution of social capital for planning and policy development strongly reflected the importance of a better understanding of social capital across a wide range of urban planning domains (Friesen, 2013a).

1.9.3 Issues of Scale in Social Capital Measurement

National and regional social trends are made visible through instruments such as the General Social Survey (GSS) so that we can learn about how people spend their time, how they perceive their neighbours, what they think of the organizations in their communities and so on. While this scale of analysis provides a general context for planners, it is much less valuable for community level design and decision-making. This is particularly true for social capital where local conditions are key drivers of the relational ecology of citizens (Bechard & Marchand, 2006).

Neighbourhood level social capital measurement requires a data collection approach that is sufficiently economical for deployment across cities at neighbourhood levels, i.e., a means of knowing what is happening in each of the hundreds of neighbourhoods in each city. Saturation surveys (outside of a census) are generally not feasible given cost and complexity but the increasing ubiquity of mobile devices that collect geographic positioning systems (GPS) data means ever-larger data sets (including large Facebook data on relationships and travel) are at least potentially available for emerging data analysis approaches (Bailey et al., 2018; Cohen, 2011; Pentland, 2014). For example, it would be expensive to collect a representative sampling at neighbourhood levels for an entire city with the additional burden of mapping the social networks of participants. Participant surveys, though common as a form of data generation, continue to face the problems that are inherent with respondent driven instruments (Bernard et al., 1984). Despite the difficulties and expense, this approach to gathering data is an important facet of social capital research to date (*Measuring Social Capital: An Australian Framework and Indicators*, 2004; Lochner et al., 1999). I use these methods as a means of exploring the potential for more effective social infrastructure measures. At small scales, behavioural data harvested from cell phones and other mobile devices processed by a spectrum machine learning techniques has been used to infer socio-spatial relational ties (Crandall et al., 2010; Eagle, 2005). This suggests new possibilities for planners trying to make use of social data in their planning and evaluation.

1.9.4 Selecting a Social Capital Measurement Tool

A Government of Canada expert report that reviewed social capital from a policy perspective considered measurement approaches at *macro* (structural including national-regional such as General Social Survey), *meso* (local communities, neighbourhoods, Census Tracts) and *micro* levels (individuals or immediate networks of individuals) (Franke, 2005). Findings suggested that more time and effort have been invested in designing and carrying out either *macro* level, international and national research programs (S. Chung et al., 2014; Garcia et al., 2008; Gesthuizen et al., 2009; van Oorschot & Arts, 2005), or highly contextual *micro* level ethnographic or site-specific research projects where behaviour of a specific group, type of individual or social unit is being examined (Boneham & Sixsmith, 2006; McPherson et al., 2006).The *meso* scale, which is particularly relevant to planners, represents an important research and policy opportunity. The meso level of social structure is strategic for improving research and policy tools that planners can use in developing strategies for city building that more fully integrate social implications in decision-making and resource allocation processes including a much better understanding of the social impact of planning decisions.

1.9.5 Social Capital and other Social Phenomena

Complex phenomena with unclear or highly diverse causal mechanisms may feature significant clusters of interacting systems and causal dynamics that cannot be understood in isolation from each other (Gunderson & Holling, 2002; T. Kuhn, 1970). This is important for planning and policy(Davis et al., 2017). In complex settings (e.g. human health and environmental conditions), scientists can often gain insight into higher order phenomenon by understanding the dynamics of subsystems of phenomena. Early social capital research and theory explored key causal factors such as trust and cultural norms (Bourdieu, 2008; J. S. Coleman, 1988; Paldam, 2000). Untangling the complex nature of these causal factors in neighbourhood social phenomena is ongoing (S. Chung et al., 2014) and reflects long traditions of deliberations that concern planning theory and complex phenomena (Jacobs, 1959; Lindblom, 1959). Research progress will be difficult given these realities.

Progress in social capital research may, as a result of these challenges, require significant timeframes (Ostrom & Ahn, 2003) and a fuller consideration of the nature of the problems encountered. Research approaches need to be designed to accommodate these dynamics including the ways in which they involve the contested relationship between natural and social sciences (Gadamer, 1960; H. Putnam, 1988; Rorty, 1998); challenges regarding the efficacy of theory in guiding research (T. Kuhn, 1970; G. Thomas, 1997); and the degree of confidence we can have in research results, particularly social science results, as true (or complete) representations of reality (Derman, 2011; Rorty, 1979). One consequence of this is to treat results as provisional rather than conclusive.

1.10 Significance for Planning Policy

The significance of social capital and spatial dynamics for planning have been noted in part above. Social structures arising from relationship networks and spatial use contribute to the physical form of our cities and communities. Planning faces an increasingly fragmented theoretical and practical working space since the fading of rational comprehensive assumptions and this difficulty is compounded by the significant range of professional, social science, political and land use issues that planning endeavors to engage and order (Fainstein & Campbell, 2011). Planning and policy are by their very nature, undertakings that operates at the intersection of a significant number of competing interests and enterprises – engineering, politics, ecology, sociology, commerce, and many others. It is critical, therefore, that research and practice enable diverse but coherent integration and in order to make progress on perennial challenges (e.g. relationship between transportation, land use, and social policy) (Irwin, 2010).

Big data systems and those that control them, however, are as likely to create new problems as they are to solve them (Borgman, 2015; Derman, 2011; O'Neil, 2016; Postman, 1993). Smart cities and big data may be inevitable but it has become clear that this is not an unmitigated good (Bell et al., 2015; Gregg,

2015). It is possible that over-reliance on technical solutions may turn out to be the downtown freeways of our time if supporting research isn't well-matched with common good practices (Derman, 2011; Friedmann, 2000; Grimmer, 2015; O'Neil, 2016).

One of the tests of the efficacy of this research is the potential it has to support generative approaches to planning. The study of complex systems dynamics has taught us that relatively simple or core "rules" that function effectively can lead to positive outcomes where constraining, bureaucratic structures fail (Innes & Booher, 2010; Ostrom, 2005; Sanders, 1998; Snowden & Boone, 2007). The balance of control and permission may be most clearly seen in settings where informal development is common (Gouverneur, 2014; Graham & McFarlane, 2015) and where a more open and experimental tone is set for urban development – e.g. tactical urbanism, guerilla urbanism, flexibility on mother-in-law suites in suburbs, and so on (Berger et al., 2014; Montgomery, 2014). Designing processes that feed a changing context back into the decision-making and building cycle can lead to the right kinds of change at the right times.

The arrangement of social structures, institutions, and cultural patterns can have long-term effects on societies: "The planning of institutions of both types – for the realization of planned change, and for subsequent control – is an important aspect of integrative planning. This means, for example, that planning for technological innovation ought to include planning for new social institutions" (Jantsch, 1972, p. 137; Schumpeter, 2011).

1.11 Conclusion

The three following chapters will describe the design, data collection and analysis for a new social capital instrument, a data set of spatial movements for 97 participants, and then compare the results of social capital data with the spatial movement data from the same participants. In the final chapter, I conclude by describing ways in which planners could improve their understanding and application of social and spatial insight gained from my study.

Chapter 2 details the development, design and data from a new social capital survey instrument. The General Social Survey (GSS) has provided significant insight about the state of Canadian society across a wide variety of themes since 1985. The GSS is in turn part of a larger International Social Survey Programme adding additional dimensions to its value for social science research (Lauer & Yodanis, 2004) and both the ISSP and the GSS are highly credible sources of social data. The GSS in Canada represents a significant body of social research spanning more than 30 years that is representative at a national scale. Given the focus on social well-being and relevant dynamics that impact collective health and function and the well-honed questions and themes utilized, the GSS was deemed to be a valuable source of themes and questions that could be used as a basis for a new, more efficient social capital survey instrument suitable for measuring social capital at neighbourhood levels. This chapter examines the rationale for question selection, development of the new Social Capital General Social Survey (SCGSS), deployment and the results of that survey with my sample.

In Chapter 3 data collection and analysis is undertaken for the same 97 people who participated in the social capital survey. Exploratory analytic methods are applied through a characterization of the aggregated GPS tracks in ten different metric dimensions. The purpose of the chapter is to identify which spatial features are the most promising for comparison with traditional demographic data such as age, years in the neighbourhood, income, and gender portion of the social capital survey. The participants are distributed among three income differentiated census tracts (median, standard deviation above and below) with common urban form. Results suggest that higher income areas travel greater extended distances while lower income has greater intensity in a smaller range, and that the innate form of the city shapes travel more than income. Higher income areas also appear to have less connection to neighbours.

The social capital and spatial data are compared in Chapter 4. An exploratory methodology is developed that can incorporate both the social and spatial data. The relationships between social and spatial data are analyzed using both supervised (linear regression with input predicting output) and unsupervised methods (random forest and principal component analysis looking at patterns and variable importance ranking). A reference data set is used to ensure the reliability of the three different types of analyses performed and is in keeping with the complex, non-linear nature of the phenomena being examined. The intent is to explore correlations between spatial movement and social capital or income levels at best and at the least, provide clues about the possibility of further work using a wider demographic range and larger data sets.

Finally, in Chapter 5 I conclude with an extension of the research design that includes evaluation of Christopher Alexander's work, *The Nature of Order*, as a bridge to practical application of complex, network oriented social phenomena to planning focused problems and challenges. This includes a review of how these insights can be applied for planning students, urban design, and land-use planning professionals. I argue that his methods and theories about systems and process interactions are a valuable source of policy insight for both planning theory and practice. Given the difficulty of some aspects of Alexander's ideas, I propose the development of an interrogative structure to access Alexander's ideas based on five orienting questions that are of practical use for policy makers and planners in the field, but which also inform new urban development theory. His intricate theories represent a development that is both rigorous and useful for policy, particularly as it concerns social capital and spatial dynamics. This increased rigor is critical for more adaptive and robust methods of complex urban theory, practice, and deployment at all levels of planning including the built environment, the analysis of social fabric, and development of effective policy research. The complex dynamics involved in this reflect "the social logic of space" (Hillier & Hanson, 1989) and includes cautions about the potential limits of inferring social realities from spatial use data, particularly the ways in which insights may be applied by prospective policy, research or community development leaders. The importance of advancing our techniques for mapping social capital includes the potential for improved governance and new platforms for collaboration, decision making and planning (Johnston, 2010).

Chapter 2: Advancing Social Capital Measurement: Using General Social Survey Cycles to Develop an Efficient Survey Instrument

2.1 Introduction

The General Social Survey (GSS) has provided significant insight about the state of Canadian society across a wide variety of themes since 1985. The GSS is in turn part of a larger International Social Survey Programme adding additional dimensions to its value for social science research (Lauer & Yodanis, 2004) and both the ISSP and the GSS are highly credible sources of social data. The GSS in Canada represents a significant body of social research spanning more than 30 years that is representative at a national scale. Given the focus on social well-being and relevant dynamics that impact collective health and function and the well-honed questions and themes utilized, the GSS was deemed to be a valuable source of themes and questions that could be used as a basis for a new, more efficient social capital survey instrument suitable for measuring social capital at neighbourhood levels. This chapter examines the rationale for question selection, development of the new Social Capital General Social Survey (SCGSS), deployment and results as part of the Social Imaging Project.

2.2 Social Capital Measurement Approaches

From an initial list of 35 different approaches to social capital measurement, six were identified based on scale of measurement suitable for neighbourhoods, degree of acceptance by academic and governmental research institutions, difficulty of application, and cross-scale comparison (Chapter 1). These measures are part of a wider measurement discussion theme within the extensive social capital literature reviewed earlier. Measurement of social change or social impact is a concern at both academic and policy levels (Berger-Schmitt, 2002; Bryant & Norris, 2002; Dean et al., 1994; L. Nguyen et al., 2015). Planners need ongoing research support to understand and apply the relevant aspects of social measurement to their own evaluations of well-being, project impact, and common good in terms of their own project evaluation (Caro et al., 2014; Friesen, 2013a). For the purposes of developing a synthetic social capital instrument out of the Canadian GSS cycles, the twin themes of social networks and trust are used as a selection and organizing strategy. These core themes align with core social capital dynamics as identified in Chapter 1.

2.3 Survey Design

Given that the cost of administering a forty-five-minute telephone survey is high, the survey developed for this study aims to identify a set of GSS cycle questions from the pool of all GSS questions. The review of social capital measurement tools and academic sources that use social capital reveals that measuring social networks and trust in some form will be critical (Benkler, 2006; Berkman, 1986; Boase & Wellman, 2004; Bourgeois & Friedkin, 2001; Richard M. Carpiano & Fitterer, 2014; Fujiwara & Kawachi, 2014; Veenstra, 2002). A new composite survey instrument based on GSS questions can therefore be valuable for ongoing research. In addition to social capital questions based on social networks and trust, demographic questions will need to be included as well to provide suitable social context.

As with the GSS, this synthetic instrument is dependent on respondent information provided to the surveyor and is not a direct behavioural measure (Bernard et al., 1984). This means the quantitative data collected is based on the perceptions of the survey respondents, e.g. If researchers ask a respondent "How many close friends do you have?" it is not possible to corroborate the answer in any formal way or to compare it to an established benchmark for that context. The value of asking the question is that variance across a population can be compared and analyzed.

Social network concepts are used in the qualitative elements of the research (participant perceptions of friends, family, acquaintances, trust in people, trust in institutions, etc.). Spatial network analysis is used in the empirical and quantitative elements of the research including the calculation of spatial statistics of each participants GPS data path. These will be analyzed in the following chapter and reflect behavioural data leading to a set of ten spatial measures and statistics as generated over a seven-day interval of time. A close complement to social ties is the phenomena of trust – where trust is absent, relational ties are much weaker, hostile, or non-existent. By measuring levels of trust in people and institutions as well as the size and scope of a given individuals social network, a relative measure of social capital levels within the survey population can be established for comparison with spatial data.

In addition to collecting demographic information about a representative national sample of Canadians, the GSS features specific themes with each cycle. Within the scope of my study, the analysis and description of themes includes 27 cycles from 1985 (Cycle 1) – 2013 (Cycle 27) as administered by Statistics Canada (Table 2-1).

Topic	1st series	2nd series	3rd series	4th series	5th series	6th series
	year (cycle)					
Health	1985 (1)	1991 (6)				
Time use	1986 (2)	1992 (7)	1998 (12)	2005 (19)	2010 (24)	
Victimization	1988 (3)	1993 (8)	1999 (13)	2004 (18)	2009 (23)	
Education, work and retirement	1989 (4)	1994 (9)	2002 (16)	2007 (21)		
Family	1990 (5)	1995 (10)	2001 (15)	2006 (20)	2007 (21)	2011 (25)
Social support and aging	1985 (1)	1990 (5)	1996 (11)	2002 (16)	2007 (21)	2012 (26)
Access to and use of ICT**	2000 (14)					
Social engagement	2003 (17)	2008 (22)	2013 (27)			
Giving, volunteering, participating	1997*	2000*	2004*	2007*	2010*	2013 (27)

Empty box = not available for a specific reference period
***Canada Survey of Giving, Volunteering and Participating (CSGVP)**
**** ICT = Information Communication Technology**
Source(s): **Statistics Canada, General Social Survey, 1985 to 2013.**

Table 2-1: Overview of Canadian GSS cycles with dates and themes ("CanSIM General Social Survey Overview," 2015).

2.4 Comparing Survey Data with Spatial Data

There is value in comparing survey-driven, respondent data with behavioural data. I have designed the research plan to include the same group of people for both the social capital survey and spatial movement data sets. The survey is primarily qualitative and depends on respondent's perception of themselves and their own self-reported values and estimates of trust and social networks. The spatial data is behavioural, requiring only that participants carry the devices per instructions. This does not suggest an opposition between qualitative and quantitative data (S. Chung et al., 2014; Creswell, 2008; Johnson & Onwuegbuzie, 2004; Mario Luis Small, 2011). The primary hypothesis is designed to further our understanding of the relationship between demographic data (income, age, social position, lifestyle habits, gender, duration of residence and so on), the nature and strength of an individual's social reality, and spatial-behavioural patterns. As has been discussed, social capital is a complex phenomenon which requires an understanding of context alongside careful measurement and an exploration of how various modes of inquiry can generate insight and understanding. If correlations between survey measures of social capital and human spatial patterns can be found, it may lead to both increased confidence about social dynamics. Quantitative behavioural data is typically generated by digital devices in significant quantity quite apart from formal surveys through smart phones and mobile devices. These very large and growing data sets can be analyzed through data-mining and statistical analysis tools suitable for comparison to a wide range of social behaviours, many of them having spatial aspects, towards a meaningful social patterns and how they change over time (Dugundji et al., 2012). Social media data from Facebook has demonstrated, for instance, the relationship between strength of friendship ties and the distance to transit nodes and transportation networks (Bailey et al., 2018, 2019). In my study, I do not depend on social media platform data but have gathered direct, original data the probes the specific interaction of social resources and movement habits in an urban area. My decision to stratify the recruitment areas to garner low-, middle- and high-income participants with common urban geographies is intended to shed light on the possibilities of generalizing social resources, which are very difficult to gather at scale, and movement data, which is becoming nearly ubiquitous. Income differentiation is a nearly ubiquitous element of social surveys of all kinds as witnessed by Canada's many decades of participation in the International Social Survey Programme (ISSP) through the General Social Survey (GSS). It is important, therefore, that efforts toward a generalized understanding of social capital by means of spatial pattern inference be structured to make income an important classification method. It is to the specific application of the GSS that I will now turn.

2.5 The Nature of GSS Data

The GSS is a collection of very reliable social data that are representative at national and some regional levels, but sample sizes are not extensive enough to provide statistical resolution at neighbourhood or census tract levels. The data are collected by respondent feedback (qualitative) on a variety of themes, which is then coded into a survey instrument that is analyzable statistically. The various themes of each cycle, given that they are social in nature, are all candidates that can be reviewed for questions that relate most closely to the specific social capital phenomena being explored here. This qualitative data will be used for the social capital baseline component of the Social Imaging Project and then compared to spatial use statistics to see if possible, correlations between spatial use and social capital levels exist.

GSS instrument structures follow a general three-tier hierarchy: Cycle – Module – Questions. The cycles identify the broad themes of a given GSS cycle. The modules are specific clusters of sub-themes while the questions further refine these themes into survey suitable questions about specific aspects of the broader themes. In addition, GSS questions may follow a Series representing longitudinal interests where a set of modules or questions are applied across multiple GSS cycles. These are not direct applications of the same instrument, so it is necessary to examine each Cycle to identify modules and then questions suitable for inclusion in this study.

2.5.1 Selecting Questions from Existing GSS Cycles

The sorting process began with a review of themes across Canadian GSS cycles 1-27. I developed a process to sort through the GSS cycles and reduce them to a set of questions directly related to social networks and trust (in people and institutions). Statistics Canada identified that a typical GSS survey can take 45-60 minutes. The difficulty of administering a survey of that length and spread of social network and trust questions being spread across various GSS cycles warranted a logical reduction of questions to a manageable form. In order to develop the Social Capital General Social Survey (SCGSS) survey instrument, a sorting process was developed to identify overlaps, unique questions and irrelevant sections pertinent to social capital (Figure 2-1).

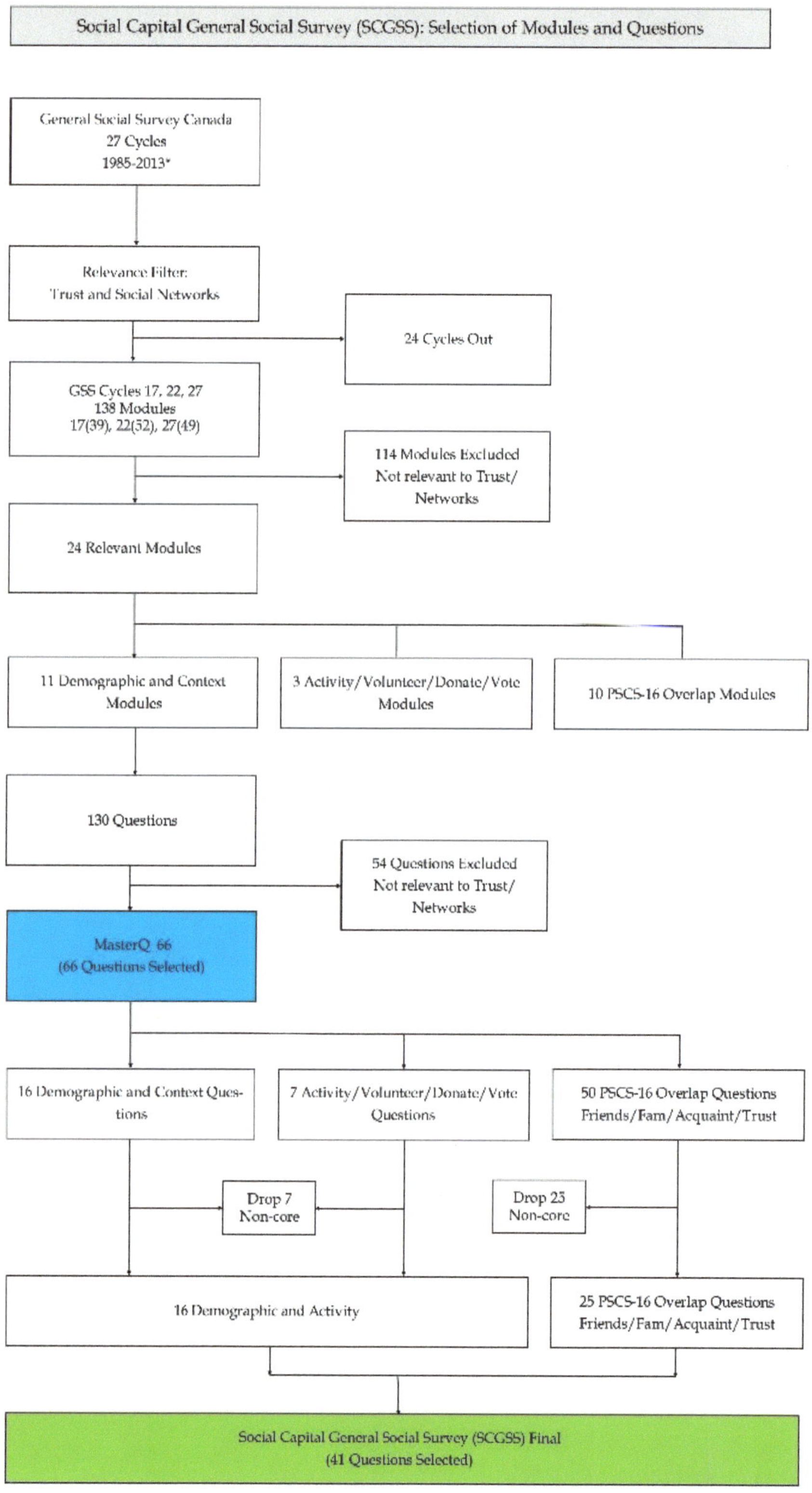

Figure 2-1 - *Logic diagram representing the process used to reduce 27 GSS cycles into the Social Capital GSS (SCGSS) instrument.*

A careful review of each cycle by theme using social networks and trust as a sorting filter led to identification of three cycles where relevance was high. These are the "Social Engagement" Cycles reflected in 1st Series (Cycle 17 - 2003), 2nd Series (Cycle 22 - 2008), 3rd Series (Cycle 27 - 2013) and 6th Series (Cycle 27). Within the Cycle 1-27 framework, this mean that 24 Cycles were excluded from further analysis given their lower relevance.

After GSS cycles 17, 22, and 27 were identified as the most directly relevant, they were sorted by sub-themes at a module level to determine which questions were suitable for inclusion. There are 138 modules across all three identified cycles. They are distributed as follows: Cycle 17 has 39 modules, Cycle 22 has 52 modules, and Cycle 27 has 49 modules. Further examination at this level revealed that 114 modules were not directly related to either social networks or trust and these were excluded from consideration. The remaining 24 modules were subdivided according to content: 11 modules concerned demographic and context questions; 3 modules related to activity, volunteering, donation practices and voting; 10 modules directly related to social networks and trust. Once these modules were identified it was possible to proceed to a review of questions.

Once a final pool of 130 GSS questions within Cycles 17, 22 and 27 were identified, 54 questions were excluded that did not directly support the demographic, activity, or social network/trust needs. It is important to note at this point the introduction of another mechanism that was used for the social capital portion of the selection process.

As noted in Chapter 1, the Personal Social Capital Scale (PSCS) is a compact survey instrument initially developed to be clear and direct enough for cross-cultural use (X. Chen et al., 2009). This required that it have as few questions as possible, that the questions be as simple as possible, and that they would focus on core social capital dynamics. The PSCS has been tested, and verified as reliable from a survey design standpoint (Archuleta & Miller, 2011; X. Chen et al., 2009; Wang et al., 2014). The PSCS has eight questions related to bonding capital - ties between and among friends – and bridging capital – ties to acquaintances or lesser known individuals or groups (*Measuring Social Capital: An Australian Framework and Indicators*, 2004; Goldstein et al., 2009; M. S. Granovetter, 1973).

With 66 questions remaining, additional evaluation was required to make the survey instrument more economical to administer. Demographic and Activity sections were reduced from 23 questions to 16.

The 50 social capital questions that remained were also reduced to a sub-set of the social networks category organized into Close Friends/Family/Acquaintances and direct questions on Trust, leaving a final list of 25 social capital questions for inclusion. These questions were then compared to the PSCS survey control to confirm relevance and distribution across PSCS question themes. A bipartite graph was prepared to provide a visual description that showed how the eight bonding capital and eight bridging capital questions of the PSCS were represented in the final 25 GSS questions (Figure *2-2*). The comparison gives us confidence that these final questions chosen for inclusion in the survey have a high degree of overlap with the verified and peer reviewed social capital survey instrument.

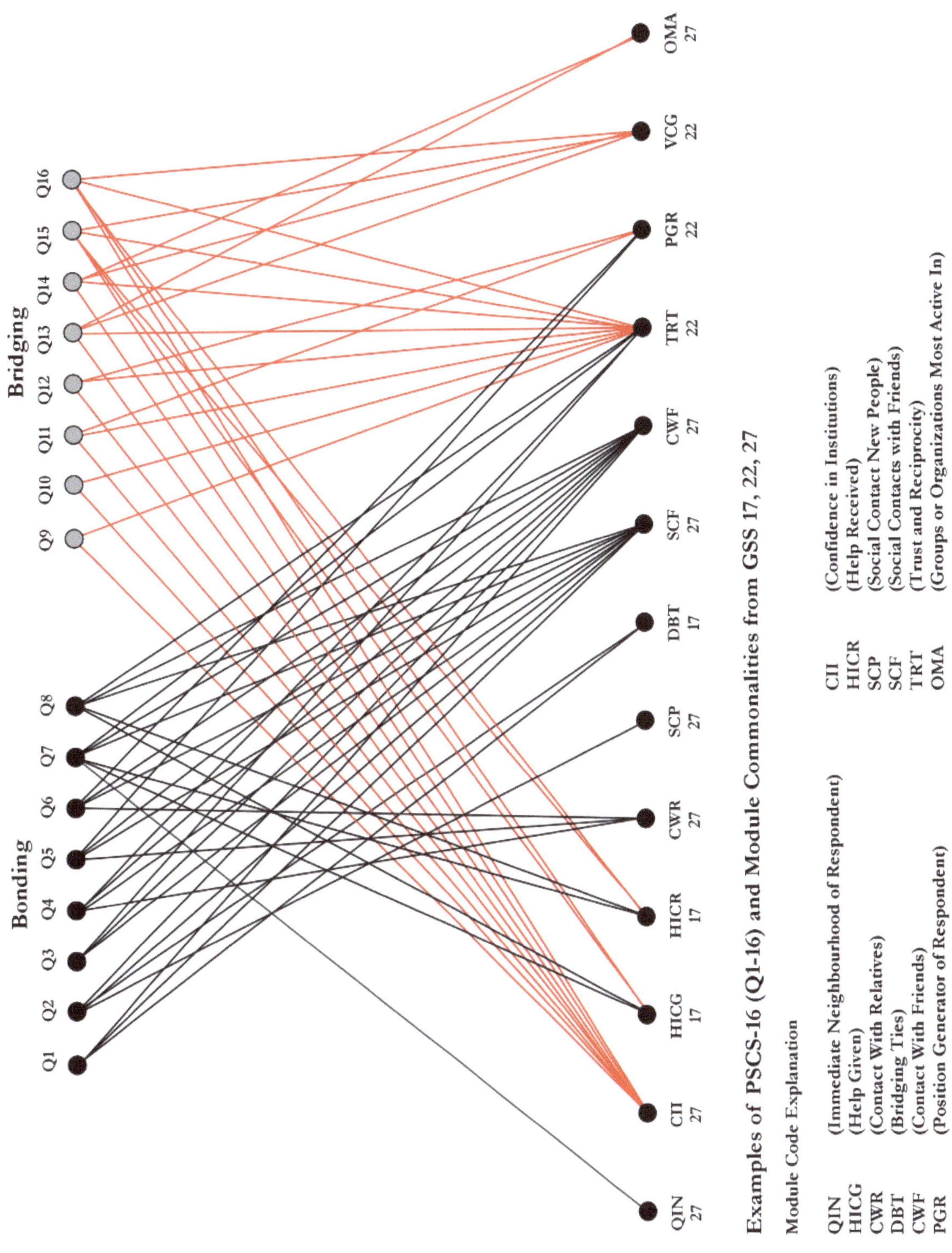

Figure 2-2 - Comparison of selected GSS modules (lower nodes) with the PSCS survey control instrument (top nodes - Q1-8 bonding capital and Q9-16 bridging capital) using social capital themes of trust and social networks. Graphic drawn by author.

The breakdown of the 25 social capital questions is organized in five sections with the following distribution of questions across those five sections (Table 2-2). The questions are concerned with: Close Friends (4 questions); Acquaintances (4 questions); Relatives (4 questions); Trust in People (5 questions); Trust in Institutions (8 questions).

Demographic and Context	Themes	Number of Questions
Social Networks		
	Close Friends	4
	Acquaintances / Other Friends	4
	Relatives	4
		12 *Social Network*
Trust		
	Trust in People	5
	Trust in Institutions	8
		13 *Trust*
		25 Social Capital

Table 2-2: Final SCGSS Question Categories

2.5.2 Identification Codes for SCGSS 41 Questions

The 41 questions used in the survey (16 Demographic and Context, 25 Social Capital) are as follows and include identifiers for Module code and Cycle. In some cases, slight modifications to the questions were required for clarity or efficiency but the principle was retained that the questions in the SCGSS reflect the form used in the respective GSS cycles. The questions codes can be identified as follow: RLR_Q100_22 means "RLR" is question theme code "Religion of Respondent" question number 100 of Cycle 22. Where the questions followed a sub-set pattern, all-caps text denotes how questions were used in the GSS, e.g., SCF_Q100_27 has sub-questions identified as CONTACT_TYPE_CLOSE_FRIENDS. These codes are included to permit tracing the SCGSS questions back to their source. In the SCGSS, however, the questions follow a simple number sequence: Q1, Q2...Q41. The questions with response options eliminated for clarity, are as follows with the Survey Cycle, module, and question identified where relevant (demographic and other data are not noted where they are in common use across GSS cycles):

2.5.2.1 *Demographic and Context (16 Questions)*

Q#	GSS Name Code	Question
Q1	AGE	**What is your age?**
Q2	GENDER	**What is your gender?**
Q3	MARITAL STATUS	**What is your marital status?**
Q4	COUNTRY	**What country were you born in?**
Q5	RLR_Q100_22	**What, if any, is your religion?**
Q6		**What is your annual income before deductions?**
Q7	EDUCATION	**What is the highest level of formal education you have completed?**
Q8	DOR_Q110_22_M	**In what type of dwelling are you now living? Is it a:**
Q9	LRD_Q10_27_M	**How long have you lived in this dwelling?**
Q10	LRN_Q10_27	**How long have you lived in this neighbourhood?**
Q11	QIN_Q50_27	(Module: Questions about the Immediate Neighbourhood of Respondent) Appears in: DOR_Q623_22 (Module: Dwelling of Respondent) **About how many people in your neighbourhood do you know well enough to ask for a favour?**
Q12	DOR_Q635_22M	(replaced "local community" with "neighbourhood") **How would you describe your sense of belonging to your neighbourhood? Would you say it is:**
Q13	MAR_Q110_27	During the past 12 months, was your main activity:
Q14	VBR_27_M	(Module: Voting By Respondent) Synthesized down to one question **Including local, provincial and federal elections, how would you describe your voting pattern?**
Q15	VCG_Q310_22_M	(Module: Volunteering and Charitable Giving) **On average, about how many hours per month do you volunteer?**
Q16	VCG_Q340_22_M	(Module: Volunteering and Charitable Giving) **In the past 12 months how would you describe amount of your donations to organizations and charities?**

2.5.2.2 *Close Friends (4 Questions)*

Q#	GSS Name Code	Question
Q17	SCF_Q100_27	(Module: Social Contacts with Friends) **How many close friends do you have, (that is, people who are not your relatives, but who you feel at ease with, can talk to about what is on your mind, or call on for help)?**
Q18	CONTACT_TYPE_CLOSE_FRIENDS	(Module: Sub of Social Contacts with Friends) **Thinking of your close friends, what is your main form of contact?**
Q19	CONTACT_FREQUENCY_CLOSE_FRIENDS	(Module: Sub of Social Contacts with Friends) Added "just enough" [CWF_Q60] **Overall, how do you feel about how often you communicate with your close friends?**
Q20	CONTACT_DISTANCE_CLOSE_FRIENDS	

(Module: Sub of Social Contacts with Friends)
~SCF_Q102
How many of your close friends live within walking distance of your home?

2.5.2.3 Acquaintances/Other Friends (4 Questions)

Q#	GSS Name Code	Question
Q21	SCF_Q110_27M	(Module: Social Contacts with Friends) **Not counting your close friends or relatives, how many other friends (or acquaintances) do you have (people you know at least on a first name basis)?**
Q22	CONTACT_TYPE_ACQAINTANCES	(Module: Sub of Social Contacts with Friends) Synthesis of: ~CWF_Q30 – Face-to-face ~CWF_Q40 - Phone ~CWF_Q45 – Text ~CWF_Q50 – Email/Internet/Social Media **Thinking of other friends and acquaintances, what is your main form of contact?**
Q23	CONTACT_FREQUENCY_ACQUAINTANCES	(Module: Sub of Social Contacts with Friends) MF_ (combined and simplified CWF_Q60_27 and CWF_Q65_27 and added "Just enough") Modified: [CWF_Q60] **Overall, how do you feel about how often you communicate with your other friends or acquaintances (excluding close friends and family)?**
Q24	CONTACT_DISTANCE_ACQUAINTANCES [CWF_Q20]	(Module: Sub of Social Contacts with Friends) **How many of your acquaintances live within walking distance of your home?**

2.5.2.4 Relatives (4 Questions)

Q#	GSS Name Code	Question
Q25	RFE_Q10_27M	(Module: Contact With Relatives) CONTACT_NUMBER_RELATIVES **Not counting the people you live with, how many relatives are you in some form of contact with?**
Q26	CONTACT_TYPE_RELATIVES	(Sub of Module: Contact With Relatives) Synthesis of: ~CWR_Q10 – Face-to-face ~CWR_Q20 - Phone ~CWR_Q25 – Text ~CWR_Q30 – Email/Internet/Social Media

Thinking of your relatives (not counting relatives you live with), what is your main form of contact?

Q27 CONTACT_FREQUENCY_RELATIVES

(Sub of Module: Contact With Relatives)

Overall, how do you feel about how often you communicate with your relatives?

CWR_Q40_27M (combined and simplified CWR_Q40_27 and CWR_Q45_27 and added "Just enough")

Q28 CONTACT_DISTANCE_RELATIVES

(Sub of Module: Contact With Relatives)

[RFE_Q20]

How many of your relatives live within walking distance of your home?

2.5.2.5 Trust in People (5 Questions)

Q#	**GSS Name Code**	**Question**
Q29	TRT_Q110_22N	(Module: Trust and Reciprocity) (Added mid-choice gradient "People can sometimes be trusted" and changed [3] to "It is better not to trust people") **Generally speaking, would you say that most people can be trusted or that you cannot be too careful in dealing with people?**
Q30	TRT_Q310_22M	(Module: Trust and Reciprocity) **How much do you trust people in your family?**
Q31	TRT_Q330_22	(Module: Trust and Reciprocity) **How much do you trust people in your neighbourhood?**
Q32	TRT_Q390_22	(Module: Trust and Reciprocity) **How much do you trust people you work with or go to school with?**
Q33	TRT_Q400_22	(Module: Trust and Reciprocity) **How much do you trust strangers?**

2.5.2.6 Trust in Institutions (8 Questions)

Q#	**GSS Name Code**	**Question**
Q34	CII_Q10_27M	(Module: Confidence in Institutions) (changed to match "trust" scales substituting "confidence") **How much confidence do you have in <u>the police and justice system</u>?**
Q35	CII_Q30_27M	(Module: Confidence in Institutions) (added "public") **How much confidence do you have in the <u>public school system</u>?**
Q36	LOCAL_GOVERNMENT	(Module: Confidence in Institutions) Modified from Federal Government to Local Government / City Council [CII_Q40] **How much confidence do you have in <u>city council</u>?**
Q37	CII_Q50_27M	(Module: Confidence in Institutions) (above + changed "major corporations" to "big business")

		How much confidence do you have in big business?
Q38	CII_Q55_27M	(Module: Confidence in Institutions) (removed "merchants and" and changed "business people" to "businesses") **How much confidence do you have in local businesses?**
Q39	REL_Q_New	(Though numerous at neighbourhood levels, GSS excludes measurement of perception of religious orgs) **How much confidence do you have in neighbourhood religious organizations?**
Q40	TRT_Q640_22M	(Module: Trust and Reciprocity) (ordering of Likert Scale reversed from original Cycle 22 use) **How much confidence do you have in the health care system?**
Q41	TRT_Q660_22M	(Module: Trust and Reciprocity) (ordering of Likert Scale reversed from original Cycle 22 use) **How much confidence do you have in the welfare system?**

2.6 Testing SCGSS

One of the benefits of using GSS instruments for the SCGSS is that GSS questions are thoroughly tested and vetted by Statistics Canada through extensive expert consultations, formal testing, refinement and field use over, in many cases, multiple cycles. In the case of repeated themes, some questions have been used repeatedly for longitudinal sampling. Response options for questions mirror the GSS options to allow for comparability with respective original cycles except in cases where there were variations from one cycle to the next or where comparability within the SCGSS could be enhanced by use of a common scale – e.g. moving from a 3-score Likert scale to a 5-scale Likert scale.

All questions were single answer and did not include open-ended responses with the exception of "Other" categories where a write-in was permitted if given by respondent. There were very slight modifications to original questions related to trust of institutions and organizations (e.g. a change from "Do you have confidence in (x)" to "Do you trust (x)"). In some cases, "Don't Know" was also added as a category where it was missing to allow for consistency across the SCGSS. I tested the instrument on 6 volunteers for clarity and possible tuning in a Hamilton context. This revealed no substantial issues. The SCGSS was tested fully a second time by Malatest, the market research firm hired to administer the SCGSS. The total number of tests is estimated at 20 voluntary participants. Malatest (hired calling firm), used a Computer Assisted Telephone Interviewing (CATI) system for administering the instrument and recording results. In some cases, where it was deemed more suitable by Malatest in consultation with a respondent who wished to send a written response, paper versions were mailed with a stamped return envelope. This was deemed necessary given the challenges of telephone response rates.

2.7 Deployment of SCGSS

Initial research design included identification of three distinct populations from which a statistically significant random sample could be drawn. These three populations represent a random sample stratified by income that share common urban form and similar geographic location in the city. This approach is much more geographically focused than the national sampling approach used by GSS cycles. One of my key research objectives is to further an understanding of neighbourhood levels of social capital and how those levels vary with resident movement in those neighborhoods across income levels.

2.7.1 Identification of Census Tracts for comparison

The Census Tracts are located in the East side of the City of Hamilton, Ontario (Canada), a city of 536,000 people located half-way between Buffalo, New York and Toronto, Ontario on the shore of Lake Ontario. Hamilton is a post-industrial city with a long history of industrial development and steel production which has in recent decades been partially replaced by health sciences research and education (Warrian, 2009, 2010). Within the City of Hamilton, prefixed in Census Tract designation with the digits 537, there are 130 Census Tracts. These are ordered based on median income of adults aged 15 and over. The distribution, as identified by a histogram, reflects a normal distribution with a slight modal-right skew (0.25) and a slightly flat and extremity extended kurtosis of -0.40 (Figure 2-3).

Mean	26756
Standard Error	507
Median	26383
Standard Deviation	5782
Sample Variance	33434852
Kurtosis	-0.40
Skewness	0.24
Range	25103
Minimum	14929
Maximum	40032
Sum	3478401
Count	130

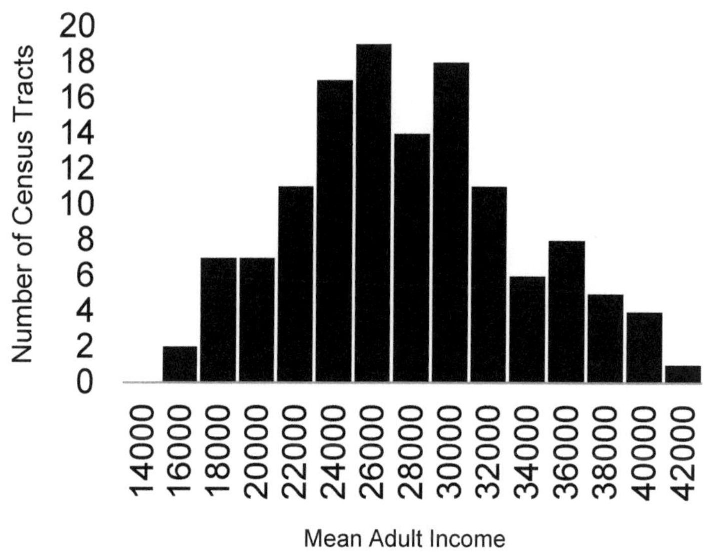

Figure 2-3: Summary statistics and histogram of income distribution for 130 Hamilton Ontario Census Tracts. Data from Statistics Canada, 2011. Histogram represents average income (ages 18-64) and number of Census Tracts in that income range.

From these 130 Census Tracts, three are identified: one median income, one a standard deviation above median, and one a standard deviation below the median. Selecting Census Tracts with average adult income at the median and closest to +/- 1 standard deviation allows testing of the central hypothesis of income impact on social capital levels: If income and social capital are correlated in some way, we would expect to see that variation at those representative income brackets. Unfortunately, these formally identified Census Tracts were in extreme locations in the city where geographic, urban form, and transportation differences were significant (e.g. Ancaster suburb, beach front along the QEW highway, semi-rural fringe)(Table 2-3**Error! Reference source not found.**).

Position	Census Tract	Income ($) yr
Median	5370073.00	26,340
One deviation above	5370131.00	32,191
One deviation below	5370053.00	20,620

Table 2-3: Census Tract identification without regard for context.

To offset a lack of common urban form and achieve a more comparable set, a search was made to determine if a better combination existed that preserved common urban form but with income differentiation. Three alternate Census Tracts were identified that met these criteria. Raw calculation for three census tracts disregarding location are as follows. In order to meet the criteria of similar urban form (housing type, street network, public transit, geographic proximity to each other) designed to minimize confounding variables, an adjusted set of three Census Tracts was identified after reviewing geographic and income variances across the 130 Census Tracts (Table 2-4).

Position	Census Tract	Income ($) yr	% difference from Non-context Census Tract
Median	5370025.00	26,671	+1.24%
One deviation above	5370030.00	31,238	-3.00%
One deviation below	5370071.00	19,178	-7.00%

Table 2-4: Context adjusted Census Tracts.

The urban form of the area encompassing these three CTs is comprised mostly of single-family detached dwellings. The center point of the three CTs lies well within a 2 kilometer radius of each other with an outer boundary comprised of both natural, street and highway features (Figure 2-4).

Figure 2-4: The white circle in the top image is a 2km radius around the center point of the three core Census Tracts. The lower map denotes the primary boundaries of the study area: Ottawa Street to the West, Barton Street to the North, the Red Hill Expressway to the East and the Escarpment to the South.

According to the 2016 Canadian Census Data, in Rosedale 89% of dwellings are single family detached (1015/1140) and in Delta 88% are single family detached (1460/1660). There are no multi-dwelling units over 5 stories in Delta, just 5 in Rosedale, but there are 700 units in apartment towers over 5 stories in McQueston. McQueston also has a much higher number of low-rise apartments, townhouses, and duplexes as can be inferred from the figure of just 44% single family detached dwellings (1200/2720) in the Census data. All three CTs have an average household size of 2.5 people but the population density does vary. Delta has the greatest density at 3874 people per square kilometer followed by McQueston at 3449 people per square kilometer. Rosedale has 1244 people per square kilometer owing to a golf course and significant green space but where residential development exists, it occurs at a density similar to both Delta and McQueston.

The McQueston neighbourhood comprised part of the original Hamilton Municipal Airport which was established in the 1920s and closed in 1951 after the current airport at Mount Hope was established. Housing in McQueston was built after 1951, including the largest social housing complex in the city which was built in the early 1960s (P. Wilson, 2009). Development in Delta (Crown Point) began between 1906-1912 but took time as the "vast sea of mud" located 1 ½ miles from the City of Hamilton didn't attract home builders even when lots were only $5 down 1906 but by 1910 more than 300 lots had sold and the building boom began (Hamilton Spectator, Staff Writer, 1936). Most of the permanent brick homes began to be built in the 1920s - part of the "great Eastward move of the City [of Hamilton]" that followed the Escarpment, met the earlier Delta/Crown Point development (Hamilton Spectator, Staff Writer, 1936), and eventually the "Adam Condy" farm was subdivided and became Rosedale between the 1940s-1960s (Figure 2-5). The most significant transportation development in the last 100 years was the opening of the Red Hill Valley Parkway. The project was in the proposal stages beginning in 1951 before it was completed in 2007. The Parkway forms the East boundary of both McQueston and Rosedale.

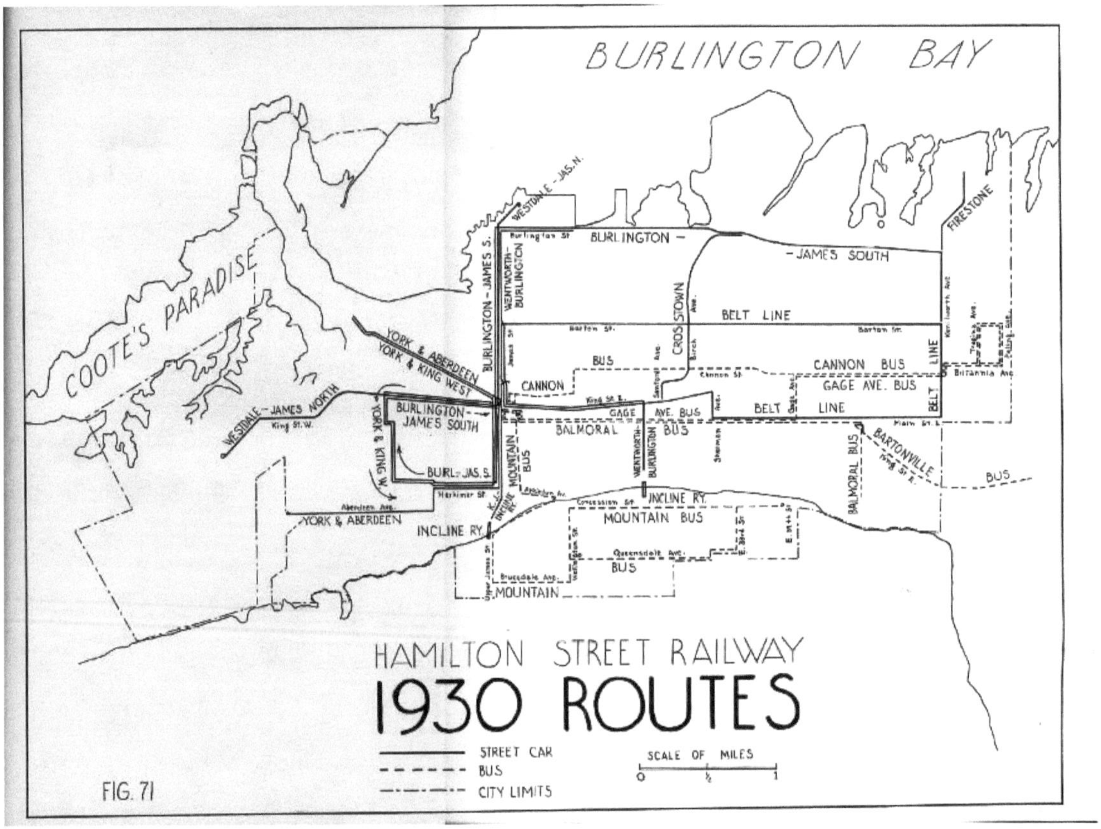

Figure 2-5 - Historical map of 1930 transit routes with "Bartonville Bus" and dashed city boundary line showing Rosedale as an undeveloped area (Hamilton Public Library Archives

2.7.2 Recruitment of participants

Participants were recruited using Random Digit Dialing for land-line phones. The study required confirmation of address which disallowed the use of cell-phone recruitment. The wider cultural trend of giving up landlines in favour of cell-phone only households increased the recruitment challenges (Table 2-5).

Neighbourhood	Eligible Population	Margin of Error	Available Sample	Target Completions	Required Raw Response Rate
Median: Rosedale (CT 25)	1840	+ 10%	775	92	11.8%
Deviation Above: Delta (CT 30)	2880	+ 10%	1,001	93	9.2%
Deviation Below: McQueston (CT 71)	4350	+ 10%	1,470	94	6.4%

Table 2-5 *Data utilized by Malatest (survey firm) Methodology Report. Population boundary is adults ages 18-64 who live in Census Tract. Location confirmation required a landline where name and address could be correlated. Confidence interval at 95%. ASDE Survey.*

Field testing of the SCGSS instrument took place in November and December 2015. The survey was formally deployed on January 11, 2016 and was in the field until March 10, 2016. After the first two weeks of calling, it became evident that it would be difficult to make the full quota for CT 25 (Rosedale) even with nearly complete call saturation. Low response rates across the other CTs were also noted. In an effort to increase the completions, neighbouring CTs near each of the target CTs were identified in keeping with income and geographic similarity as a means of increasing population sizes (Table 2-6). The survey house also used mail-outs to telephone recruited participants where that preference was identified in the recruitment calls.

Neighbourhood	Eligible Population	Available Sample	Target Completions	Required Raw Response Rate
McQueston(CT 71)	4350	1,470	92	4%
McQueston (CT 59)	3173	668		
McQueston (CT 57)	3070	660		
McQueston (CT 69)	883	217		
McQueston (CT 53)	3096	696		
McQueston (CT 58)	2361	529		
Rosedale (CT 25)	1840	775	94	3%
Rosedale (CT 28)	2849	872		
Rosedale (CT 55)	2943	728		
Rosedale (CT 54)	2732	581		
Rosedale (CT 56)	3266	880		
Delta (CT 30)	2880	1,001	91	6%
Delta (CT 31)	2098	533		
Delta (CT 29)	4335	1206		
Delta (CT 27)	1064	318		
Delta (CT 70)	2207	515		

Table 2-6 *Secondary Census Tract options in order of importance within each neighbourhood.*

The expanded range of survey territory increased the population size that recruiters could make use of to increase the number of completions (Figure *2-6*). Rosedale's expanded CTs included CT 28, CT 56, CT 55 and CT 54. McQueston's expansion included CT 57 and CT 59. Delta's expansion included CT 29 and CT 31.

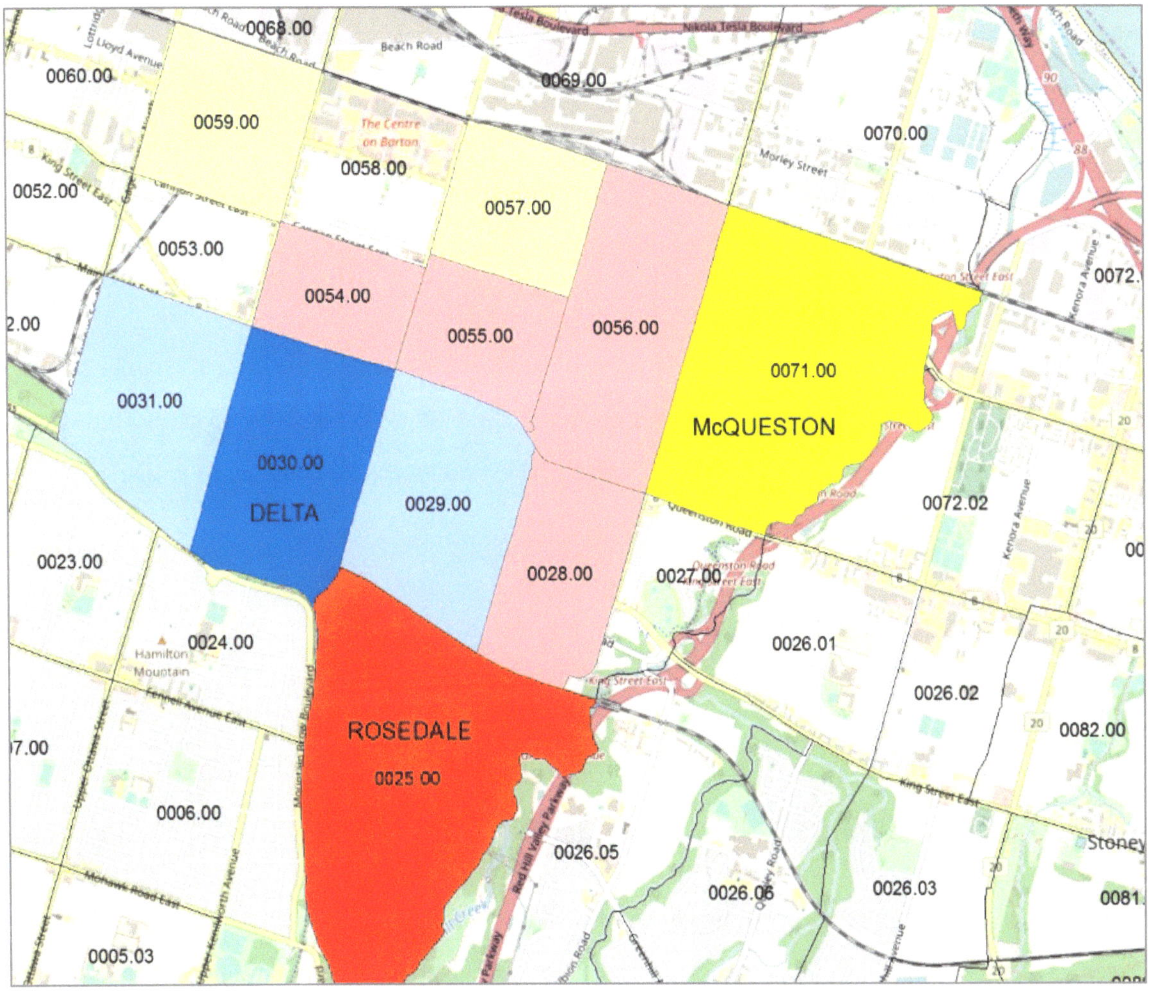

Figure 2-6 - Core Census Tracts in solid colours, secondary tracts share common colour but lighter. Priority established by common income averages + proximity. Census tract boundaries and fill done with ArcGIS. Basemap provided by Open StreetMap (and) contributors.

2.7.3 Data Collection

During data collection, 6099 calls were made to the geographic areas identified above leading to 153 study participants. Of those 153 recruited participants, 97 completed all of the required elements correctly: the SCGSS survey, travel data GPS collection, and travel diary.

The response rate was 2.5% and the refusal rate was 29% across a total of 6099 total call attempts and 473 hours invested in survey recruitment calling. Even with clear identification of the purpose and public good nature of the study, it seemed that given the requirement to complete both the telephone survey and agree to carry an electronic device, the perception of difficulty exceeded the interest. The target yield was ~90 completes per Census Tract (n=270) to meet a generalizability requirement. Given time, budget, and population sizes in the face of the low response rate, the number of completes after vetting for data problems, incomplete surveys, and malfunctioning devices was well below this number. The results, therefore, will be of comparative value within the participant pool and generalizable to a certain area of the City of Hamilton but cannot be statistically extended beyond those limits.

2.7.4 Data Handling Process

The full SCGSS data frame consists of 41 survey variables, one ID variable, and 10 spatial variables for 97 participants. Utilizing the data structure feature in R, a data frame was generated with variable naming conventions compatible with R. The spatial variables will be considered in the following chapter.

The general framework used for data preparation and analysis are as follows:

1. *Raw data;*
5. *Technically correct data;*
6. *Consistent data;*
7. *Statistical results;*
8. *Formatted output. (de Jonge & van der Loo, 2013)*

Elements 1-3 above represent data cleaning where the raw data provided by the contracted survey firm is reviewed, decisions made about changes to correct clear and obvious errors (missing values, data entry errors where values are beyond survey range, etc.) and finally adjustments made to render that data suitable for use in an R environment (integer, numeric, and factor values corresponding to survey questions). Elements 4-5 above represent the data modeling decisions leading to the production of tables, graphs and charts that reflect information useful for evaluating the central and secondary hypotheses I have formulated.

The data from the telephone survey and mail in diary and GPS data collection was provided to me from Malatest. The data was in Excel format with the survey data and GPS data in separate files. Data for the 41 questions were reviewed to determine completeness, missing values, unusual values (e.g., 1-5 response range but number was 8, etc.). For graphing individual variables, responses such as "refusal" or "do not know" were utilized in their text format. A numeric format was used for regression and machine learning analysis where those values could not be processed in text form such as "88" (refusal) or "99" (do not know).

2.7.5 Geographic Distribution of Completed Surveys

In order to plot the locations of each of the survey respondents, the street addresses were geocoded and loaded in Google maps as addresses only – no data or personal information was included – to review a reasonable distribution according to the Census Tract boundary mapping strategy (visual display of individual participant locations, even after random offsetting before plotting, was not permitted per Human Research Ethics, University of Waterloo). As noted, response dynamics required an increase in population boundaries. In the case of Delta, the cluster stayed very nearly within the formal Census Tract boundary. In the case of Rosedale, and to some extent McQueston, secondary Census Tracts had to be used. Rosedale (Mid, n=28).

The Rosedale Census Tract was by far the most unresponsive. This was not expected. There were no obvious contextual or demographic differences that would have suggested a response rate that was different than the other Census Tracts. This did not occur and so responses include Census Tracts 55, 54, 56 called in that order while attempting to achieve survey yield targets (94). Only 29 surveys were completed in Rosedale within the research budget and time limitations. Final analysis identified one participant file that was incomplete which lowered the final number to 28.

The McQueston Census Tract was the largest in terms of population and residential usage. Calling quotas were also problematic in this Census Tract. A hoped-for yield rate of 92 completes resulted in just 32 within time and budget limitations. In order to get to 32 completes, Census Tract 59 was opened up for calling. It was expected that a potentially lower call response rate would be seen here given greater ethnic diversity but that the size of the Census Tract would offset this limit. Instead, a lower response rate was encountered throughout the Census Tract. Census Tract 59 proved more responsive and bolstered completion numbers.

The highest income Census Tract was Delta and response rates in that area were the best in the study. Yield rates remained well below expected and additional population was needed for the sample leading to seven additional completes from the neighbouring Census Tract 31 which lies to the West of Ottawa Street. During data analysis, two additional participants had incomplete data files and were excluded, dropping the final number of completed surveys to 37.

2.8 Results of SCGSS

There are a number of notable features of the survey population. Specific variations among the three identified sub-regions will be explored further below. The gender of respondents was 61% female and 39% male with no other categories identified. Most lived with a partner either married (47%) or living common law (12%) with a notable segment identifying as single and never married (20%) (Figure *2-7*).

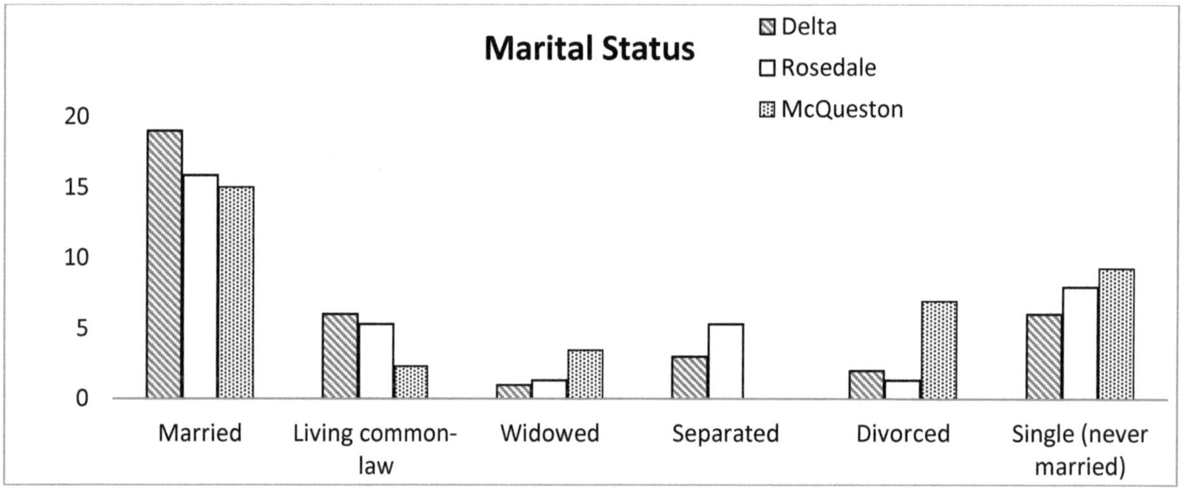

Figure 2-7 - Marital status across three Census Tracts (weighted by count).

Only 3% of respondents were born outside of Canada (none were born in the United States) and 74% had lived in their current home more than ten years. There was little movement within the neighbourhood given that those figures and dwelling tenure matched very closely. The dwelling type of 86% of respondents was single- or semi-detached houses. This reflects a stable, long-tenure population across all three Census Tracts as represented by the participants. McQueston has a slightly higher degree of diversity with representation of 2-3 participants in semi-detached, town/row house and 4 participants in apartments over 6 floors. Outside of these differences, single detached housing types, population densities, street and arterial roadways and the mix of residential and commercial properties are common across the Census Tracts surveyed. More specific details about the individual areas are reflected in the results below. Overall, this reflects contextual confirmation for similarity of urban form across the three Census Tracts chosen for the study (Figure *2-8*).

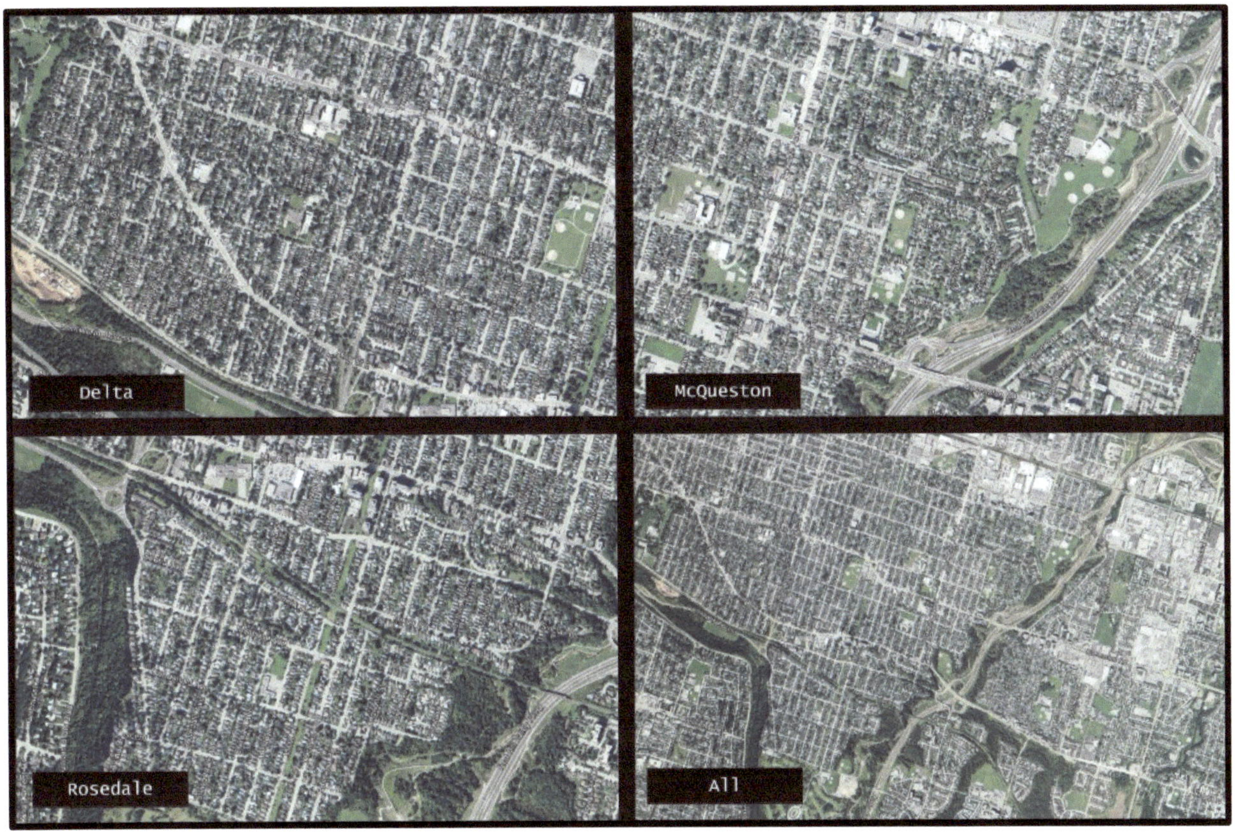

Figure 2-8 - Google Earth images of Census Tracts reflecting basic urban forms. Google Earth tours of each of the three core Census Tracts are also available.

Rosedale https://www.dropbox.com/s/53vg4l1hdebl1r5/CT%2025.kmz?dl=0
Delta https://www.dropbox.com/s/gvshn4mzmemyfbm/CT%2030.kmz?dl=0,
McQueston https://www.dropbox.com/s/3tb84gvqv2gw80l/CT%2071.kmz?dl=0

Weighted average income of the 86 participants who answered the question was $44,884 (Table 2-7).

Frequency	Income ($)	(Freq) x(Income)
5	2500	12500
12	10000	120000
15	22500	337500
12	35000	420000
9	45000	405000
10	55000	555000
10	70000	700000
9	90000	810000
4	125000	500000
		Mean = $44884

Table 2-7 *Weighted income distributions across all Census Tracts where "Income" question was not refused.*

Income shows more interesting variation when each Census Tract is considered on its own. The mean of Rosedale (Med) is in the $40,000-$49,999 bracket, the mean of Delta(plus one standard deviation) is in the $50,000-$59,999) bracket, and the mean of McQueston (minus one standard deviation) is $15,000-$29,999) which does follow the expected income based on the initial Census Tract selection process. When each bracket is compared by Census Tract, however, we can see greater variation within the segments (Figure *2-9*).

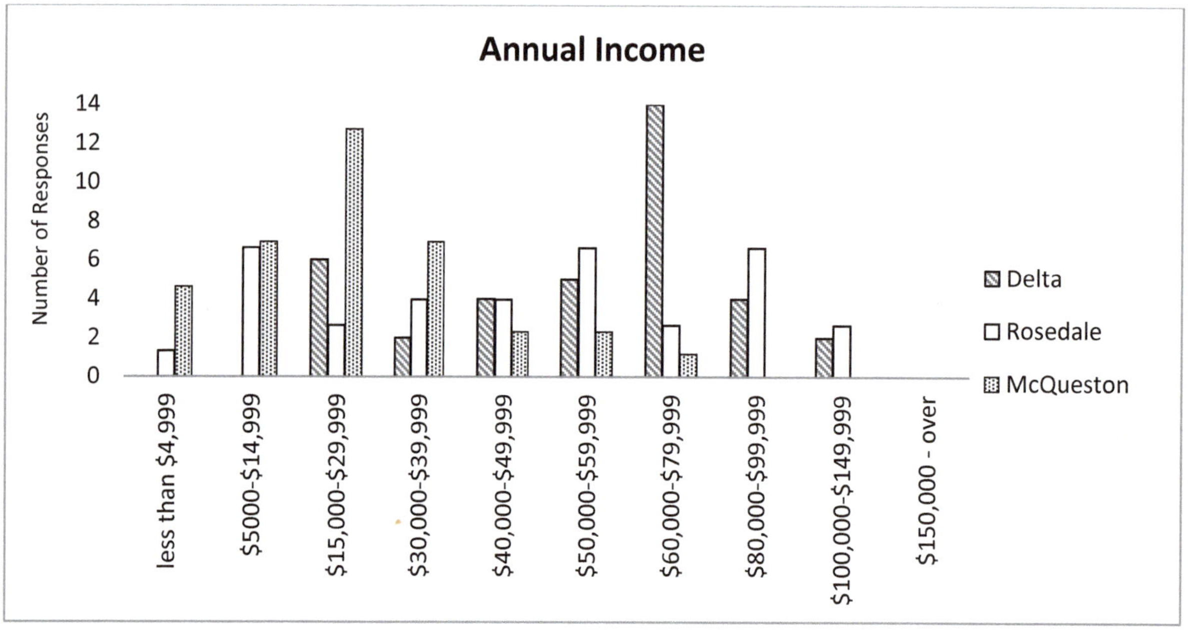

Figure 2-9 - Income distributions by Census Tract (weighted by count). Y axis = number of participants in each category. Missing bars reflect no participants in that particular bracket.

McQueston is well-represented in the lower half of the brackets with a notably higher number of residents in the $15,000-$29,999 bracket along with representation in the lower two brackets and no representation in the top two brackets. Delta is nearly the inverse by pattern with a significant spike in the $60,000-$79,999 bracket and no representation in the lower two brackets. There is a slight anomaly in the $15,000-$29,999 bracket for Delta. This confirms that the income-differentiated objectives in choosing CTs is reflected in the participant data. Extending telephone recruitment to neighbouring CTs with similar income did not impact the income differentiation objective.

The average age of all participants is 50 years while the average age of Rosedale is 48 years, Delta is 48.3 years, and McQueston is 53 years (Figure *2-10*). Religious and educational patterns reflect that 31% identify as atheists or agnostics, 62% with some form of Christianity and 64% have attained some form of education above a High School diploma.

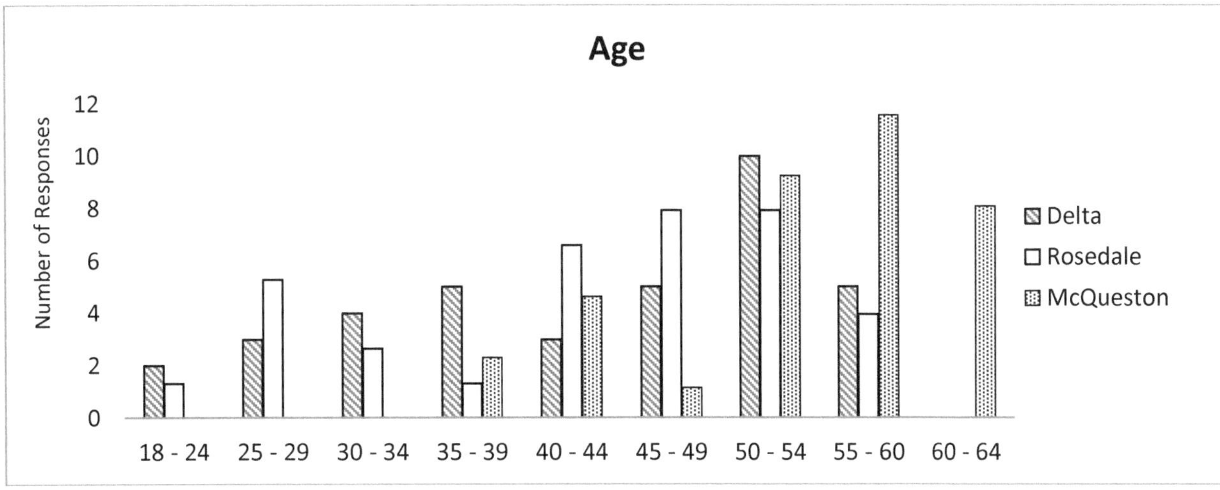

Figure 2-10 - Age across three Census Tracts (weighted by count).

2.8.1 Context Analysis

The SCGSS instrument has a second sub-theme within Demographics that is intended to provide a general sense of how people perceive and interact with their immediate neighbourhood. These questions queried sense of belonging, asking for help, voting, volunteering habits and primary activities. Across all three Census Tracts, only 3% of people felt they didn't know anyone in their neighbourhood well enough to ask for help or a favour while 49% said they knew 1-5 people and an additional 48% they knew more than six people they would be willing to ask for help. This is reflected in the question about neighbourhood belonging where 79% of people across all Census Tracts said their sense of belonging was either Very Strong or Somewhat Strong. Neighbourhood ties and sense of belonging appear to complement each other. The main activity of respondents over the past 12 months was either working (67%) or retired (16%) with the balance divided between looking for work, school, caring for children, household work, or disabled (6%).

As in the Demographic section, specific and interesting variations show up when a more granular view is taken. When asked how many neighbours they knew that they could ask a favour from, it is notable that Delta had a significant number of "None" responses – more than any other category – while showing no participants who knew more than 10 neighbours they could ask for a favour (Figure *2-11*).

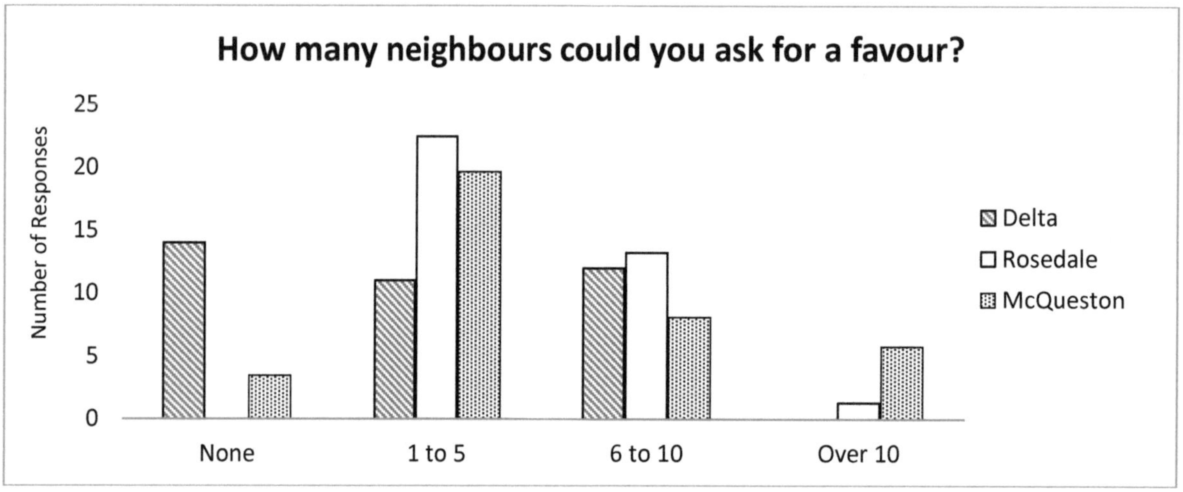

Figure 2-11 - Comparison of general neighbourhood network across all three Census Tracts (weighted by count).

Rosedale was well represented in the "1-5" bracket while McQueston shows a tendency to larger networks of neighbourhood support. Within this comparison, Delta has the weakest neighbourhood tie pattern with Rosedale showing a stronger small network and McQueston a stronger large network pattern.

Neighbourhood ties presents an interesting comparison with "sense of belonging". As noted above, overall belonging is high across the Census Tracts and each of the three Census Tracts share similar perceptions about belonging. Delta has the greatest number of participants in the "somewhat strong" bracket alongside McQueston and Rosedale. All have a similar number of "very strong" participants and the balance of categories is less significant overall (Figure *2-12*).

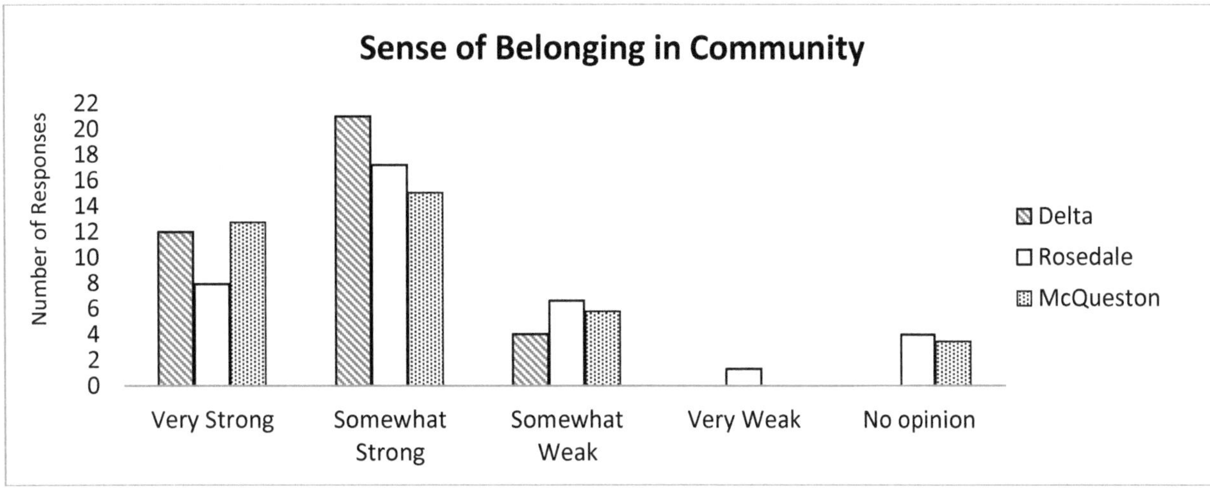

Figure 2-12 - Responses to question about how much respondent felt they belonged in their community (weighted by count).

Nearly all respondents across all three Census Tracts cite work as their primary activity and when queried about how often they vote at Municipal, Provincial and Federal levels, with few exceptions they chose "always" as a response with McQueston showing only a slight shift to less active voting. Other forms of civic engagement such as volunteering and donating reflect that most people don't volunteer any hours per month and consider their donations to charities as "modest" or "none". Delta's respondents were far more likely to consider their charity donations as "average" while those who said they gave nothing over the past 12 months were most numerous in McQueston (Figure *2-13*).

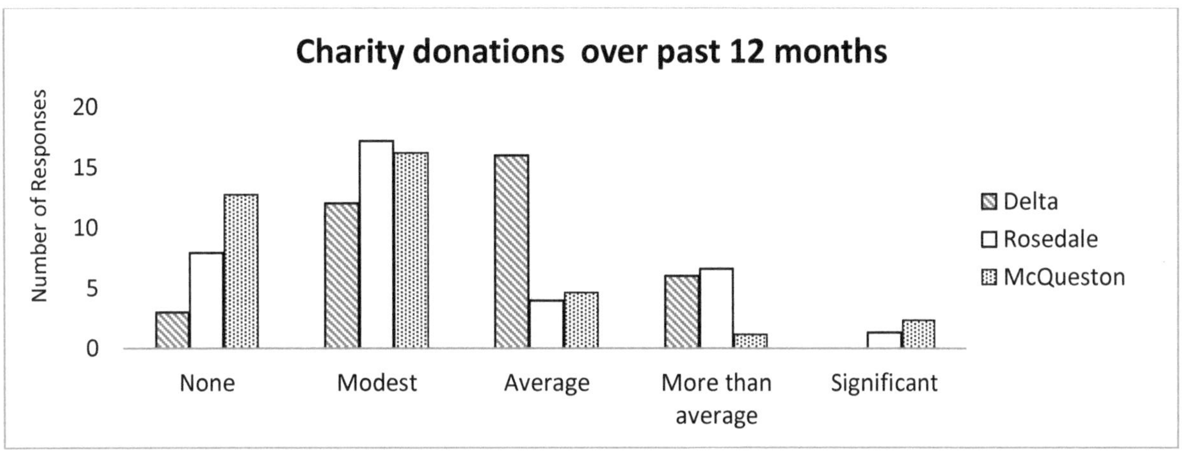

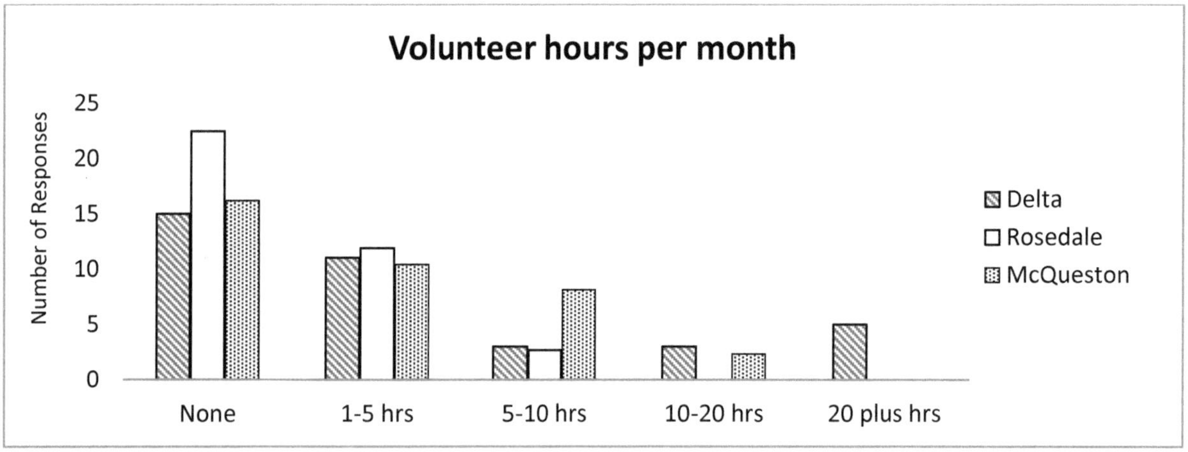

Figure 2-13 - Comparison of three Census Tracts and perceptions of charitable donations and volunteering (weighted by count).

2.8.2 Social Network Characteristics

The following section consists of participant responses to questions about both their social networks – close friends, relatives (not living with them) and acquaintances (people you might greet or know by first name only). It also explores how much participants trust people generally and their friends, family and acquaintances more specifically. Various organizations and institutions are also more or less trusted by people and their perceptions reflect a wider sense of how the individual sees themselves in the wider relational web of their community.

A comparison about the size of social networks across all three Census Tracts reveals a dynamic landscape. Most people have 3-5 close friends with McQueston trending toward a larger circle and Delta trending toward a smaller circle (Figure 2-14).

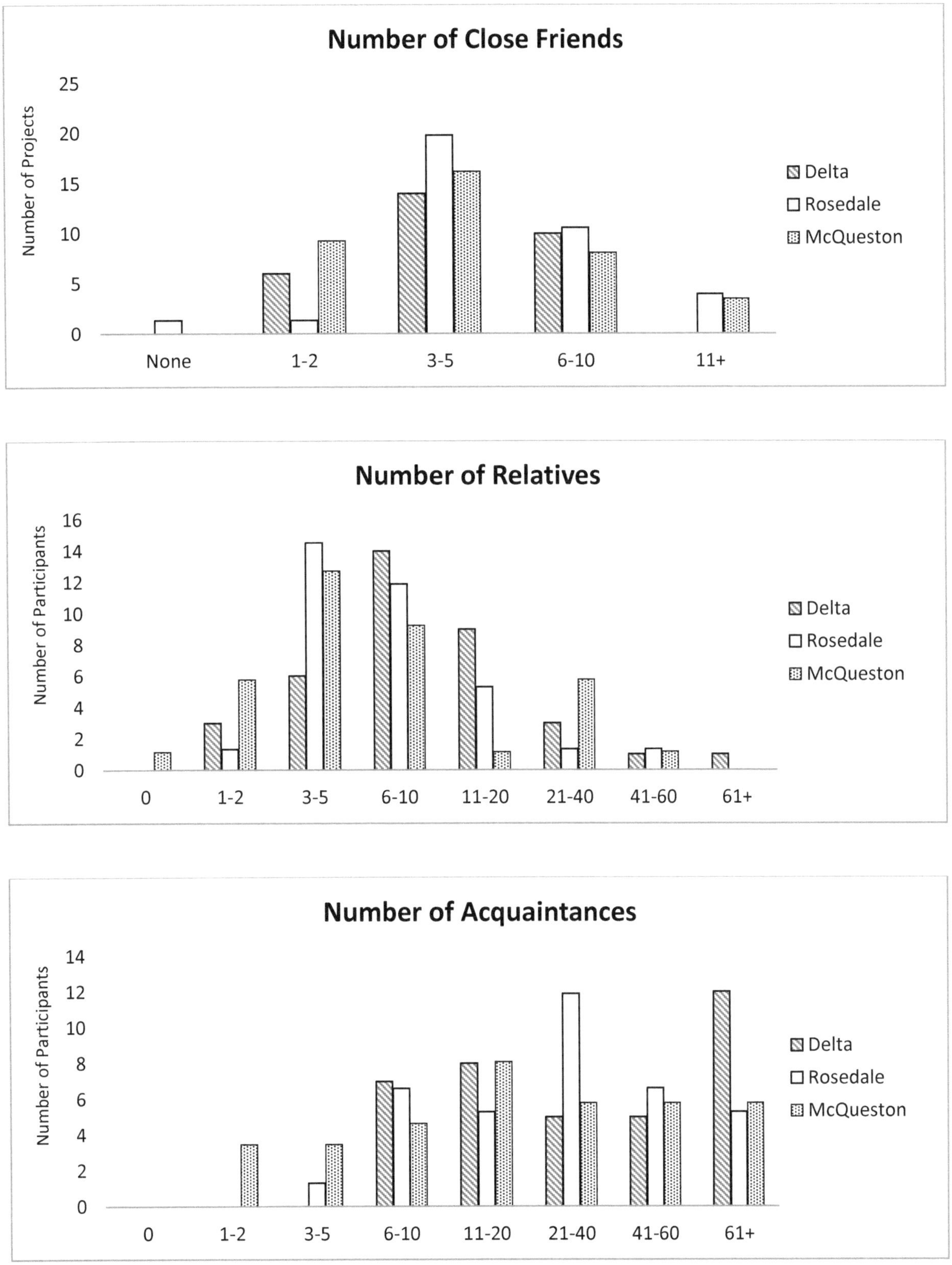

Figure 2-14 - Comparison of social network size across the three Census Tracts (weighted by count).

Rosedale reflects the highest 3-5 bracket for close friends. The number of relatives is much larger overall. Rosedale has little representation below the 3-5 bracket while McQueston shows the most even distribution across all categories. Acquaintances is the flattest of all the graphs with notable peaks in Rosedale for the 21-40 bracket and for Delta in the 61+ bracket. How we interact with people is an important aspect of socio-spatial interactions (Crandall et al., 2010; Dugundji et al., 2012; Onnela et al., 2011).

Querying participants about their mode of contact with friends, family and acquaintances provides insight about the nature of their socio-spatial landscapes. There are three overall patterns in the mode of contact responses. First, close friend contacts make use of all four modes of engagement more evenly than do acquaintance or relative networks (Figure 2-15).

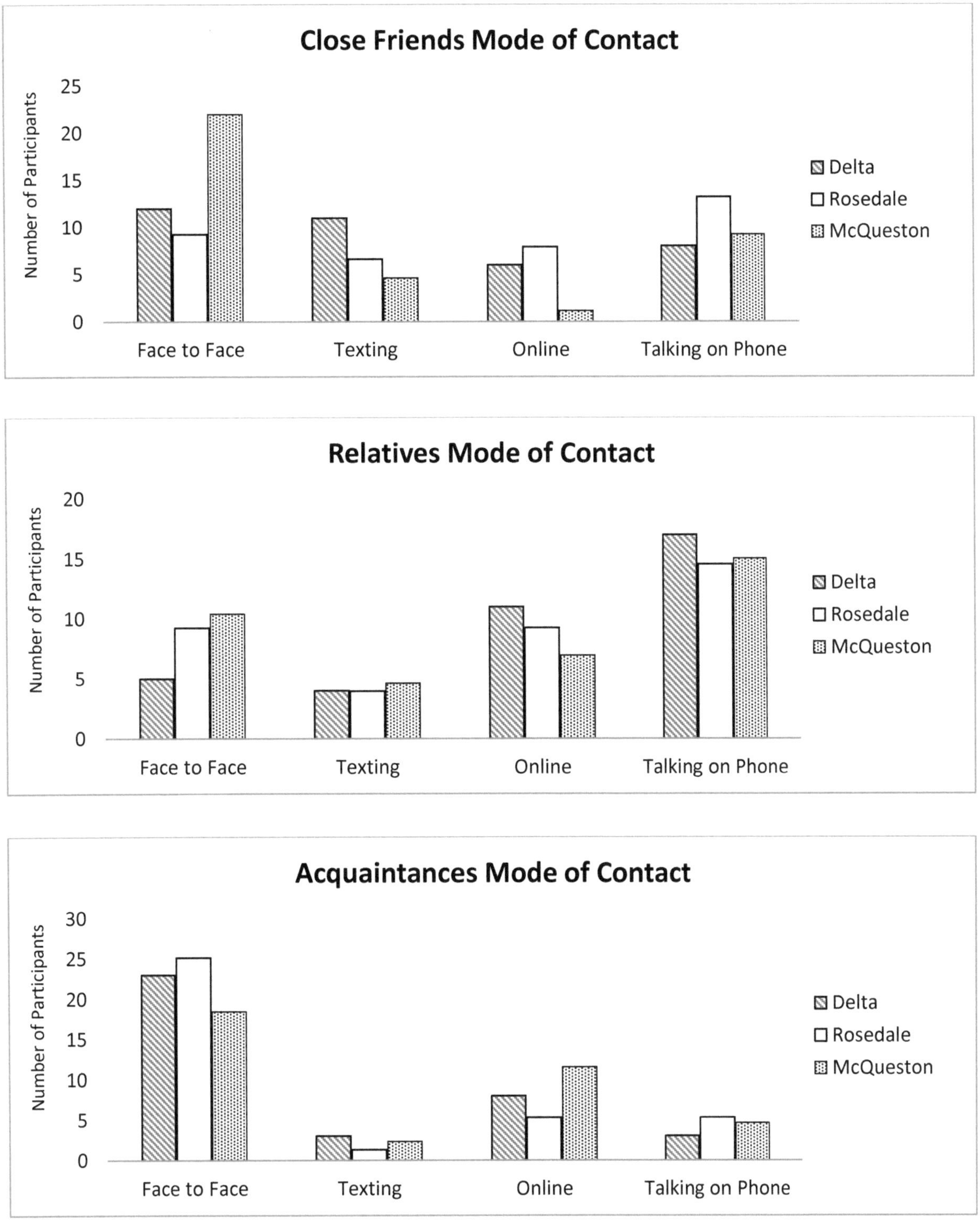

Figure 2-15 - Modes of contact for various social groups across all three Census Tracts (weighted by count).

Second, relatives are most often contacted by telephone, suggesting geographic distance of that network. Third, the mode of contact that is dominant for acquaintances is face-to-face. This likely reflects that people we know in passing are not usually well known enough for telephone conversations or other more personal digital forms of interaction and the primary point of contact is more incidental and in person. It is possible that geographically proximate relationships like this may not necessarily be significant in terms of relational intensity but when they do happen, they happen in person rather than via digital forms of interactions (texting, social media). This makes sense when you consider that people may know a local vendor (gas station, hairdresser, grocery store clerk) only by seeing them in person at a store or shop. Variations of these patterns are also present in the data. In McQueston, close friends is significantly more likely to be face-to-face than in any of the other two Census Tracts. Rosedale trends toward a greater use of telephone for friend contact than the other two Census Tracts. Delta's participants are more the most likely to text close friends – more than twice as likely as participants in McQueston.

There is increasing evidence that social isolation is increasing despite significantly enlarged communications technologies and opportunities (Holt-Lunstad et al., 2010; Marche, 2012; McPherson et al., 2006; Pantell et al., 2013; Pinker, 2014). Given this overall trend, a question was included that asked participants to reflect on the frequency of their contact with friends, relatives and acquaintances and if they were satisfied with that contact. Across all three Census Tracts and all three relational levels, the "just enough" bracket was dominant (Figure 2-16).

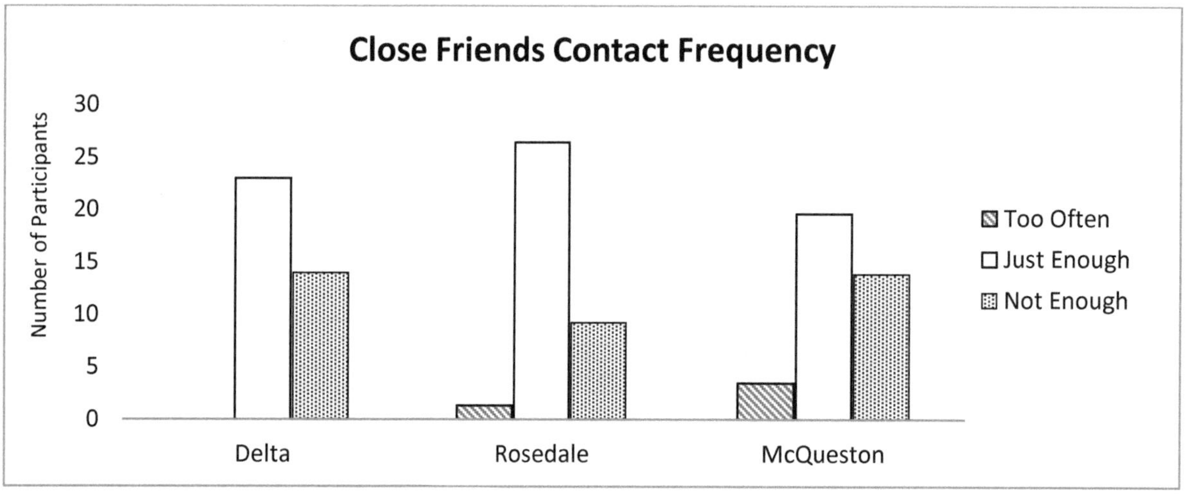

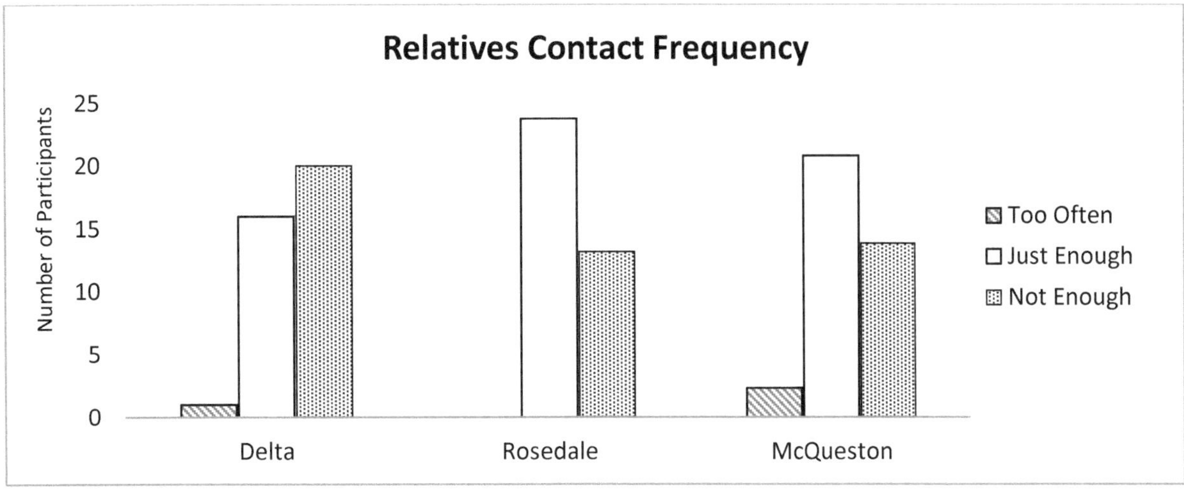

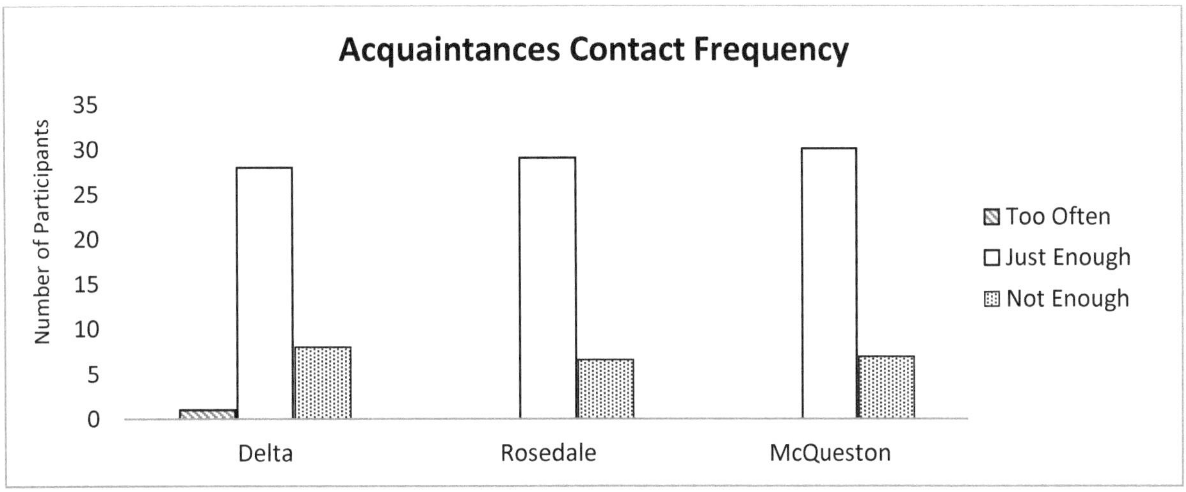

Figure 2-16 - Participants feelings about the frequency of contact with friends, family and acquaintances compared across all three Census Tracts (weighted by count).

This suggests that overall, most participants have close friends, relatives and acquaintances and are satisfied with how often they interact. It would appear that there isn't a social isolation crisis in East Hamilton if these indicators are accepted as reflective of the area. Within the overall satisfaction, however, there are some notable variations. There were only very minor signals from Rosedale and McQueston that some participants felt there was too much contact from close friends. The only instance across all three categories where "Not Enough" represented more people than "Just Enough" was in Delta where frequency of contact with relatives did not meet demand for participants. Contact frequency impressions for acquaintances was nearly identical across all three Census Tracts suggesting what is logical – that acquaintances may be the least dynamic social interaction space and the least subject to variation as a result.

Related to frequency of contact is a spatial question that probes the degree to which these three relational circles are geographically proximate to the participants. Walkability has received considerable and growing attention in recent years and there was an interest in discovering how near friends, relatives and acquaintances are. The representation of a localized and walkable network of relationships shows a definite pattern: one or two close friends may be in walkable distance but almost no relatives are for participants in all three Census Tracts (Figure 2-17).

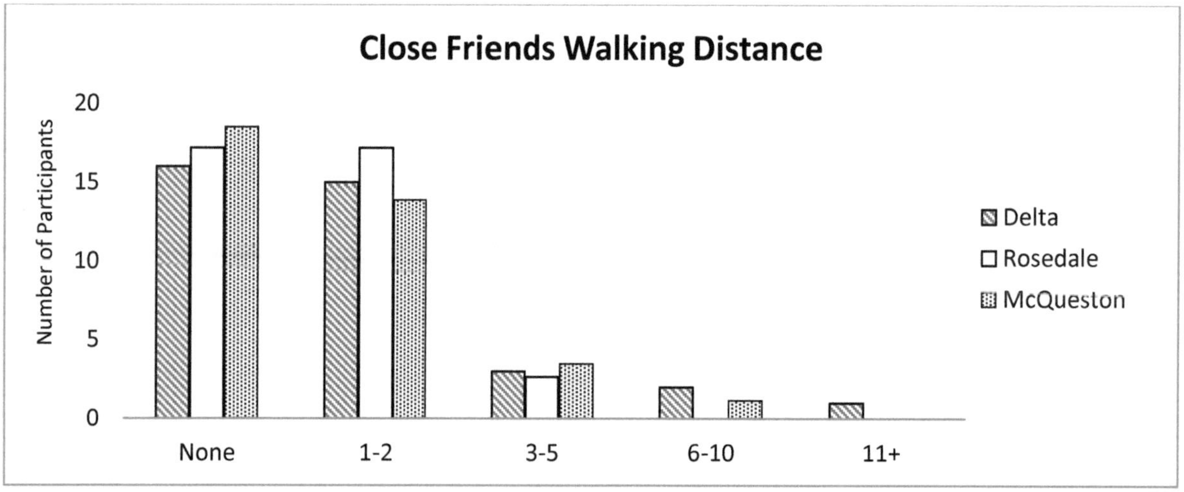

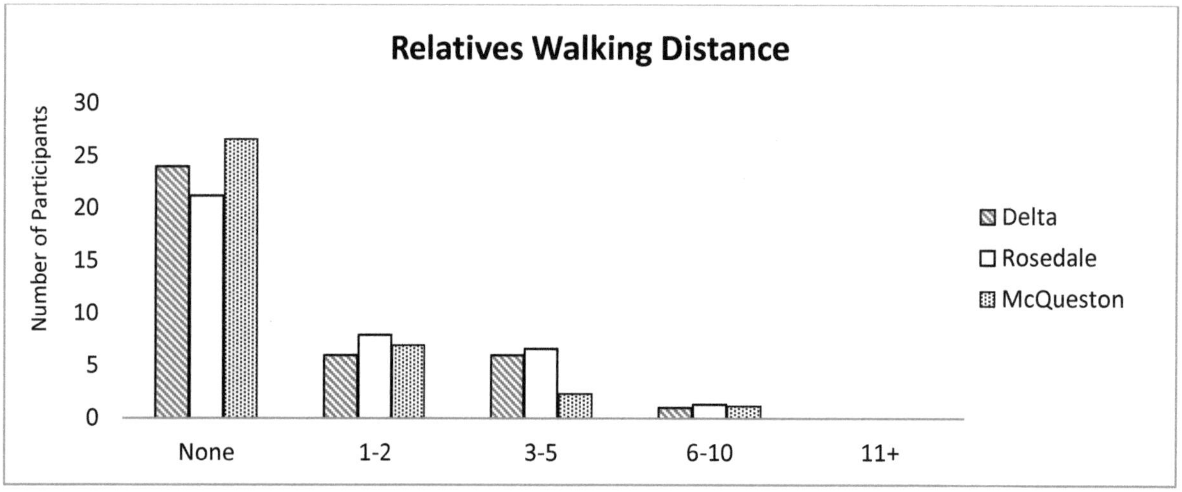

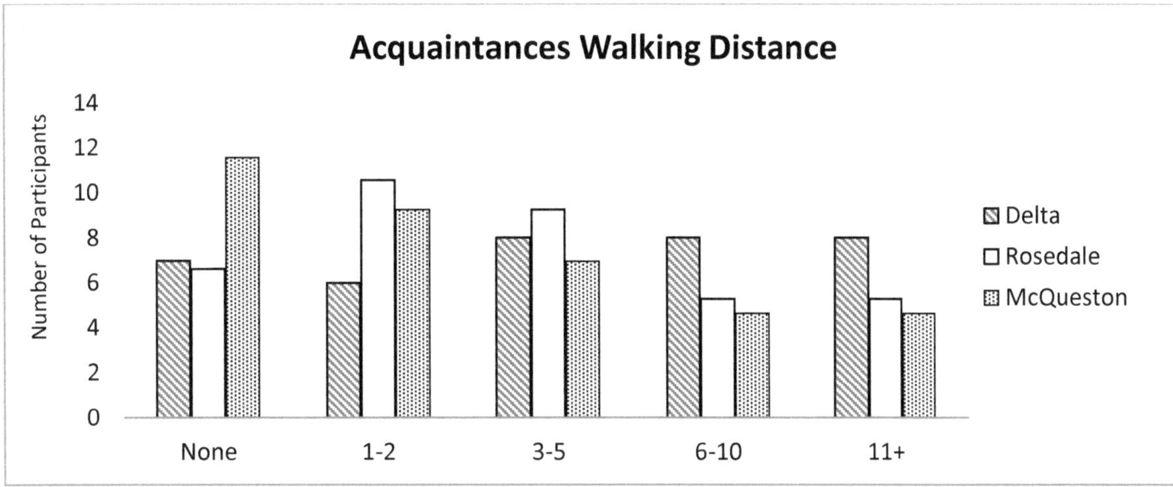

Figure 2-17 - Walking distance comparison across all three Census Tracts (weighted by count).

Uniformity continues with acquaintances but in a different form. Nearly every network size is represented evenly in each Census Tract with the overall number of acquaintances in walking distance being significantly greater than the other two categories. Given that face to face is a significant mode of contact for acquaintances, it seems probable that those interactions are happening in and around the neighbourhood of participants. We tend to have fewer face-to-face acquaintance contacts if we are commuting by car though that number could increase for transit users. Transit was not included in this study. There appears to be a strong local network of acquaintances for participants regardless of which Census Tract they live in.

2.8.3 Trust

The second aspect of social capital that is investigated by the SCGSS is trust. In turn there are two different types of trust that are probed: trust in various individuals and trust in institutions. Social capital is a network phenomenon with individuals as the constituent component and groups of individuals as a resulting secondary component. Trust is a critical feature of all social relationships and, while it is a perceptive and dynamic phenomenon, relative values pertaining to more or less can be used to gain an understanding of who or what people have confidence in.

The first two comparisons are between a general sense of trust in people and trust in strangers. The first type of trust explores trust as an aggregate of all relationships while the second is a more specific category that considers an individual who is not known but may be encountered in a given community.

The most notable similarities are that Rosedale has a similar level of moderate trust for both strangers and people in general while (Figure 2-18).

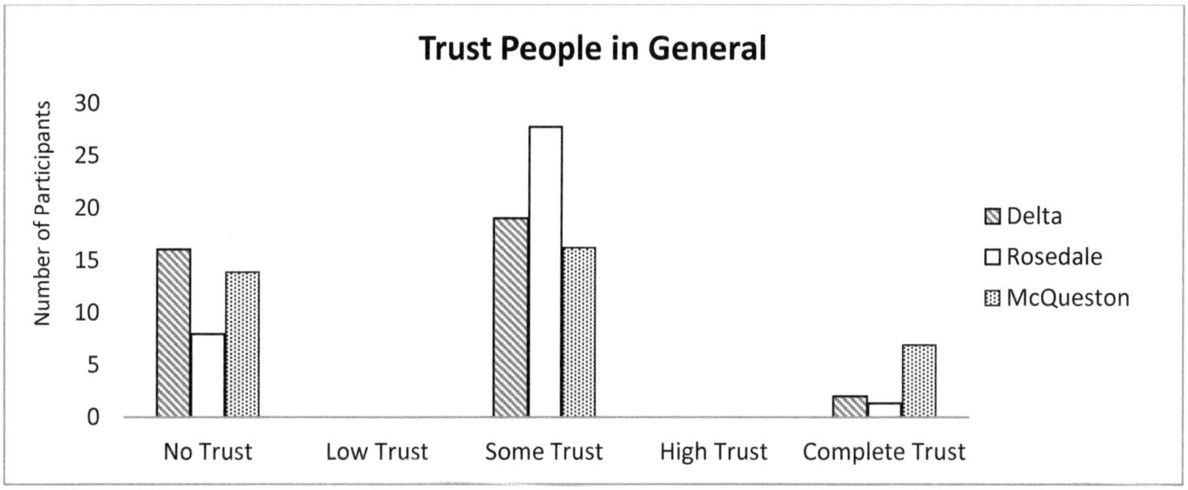

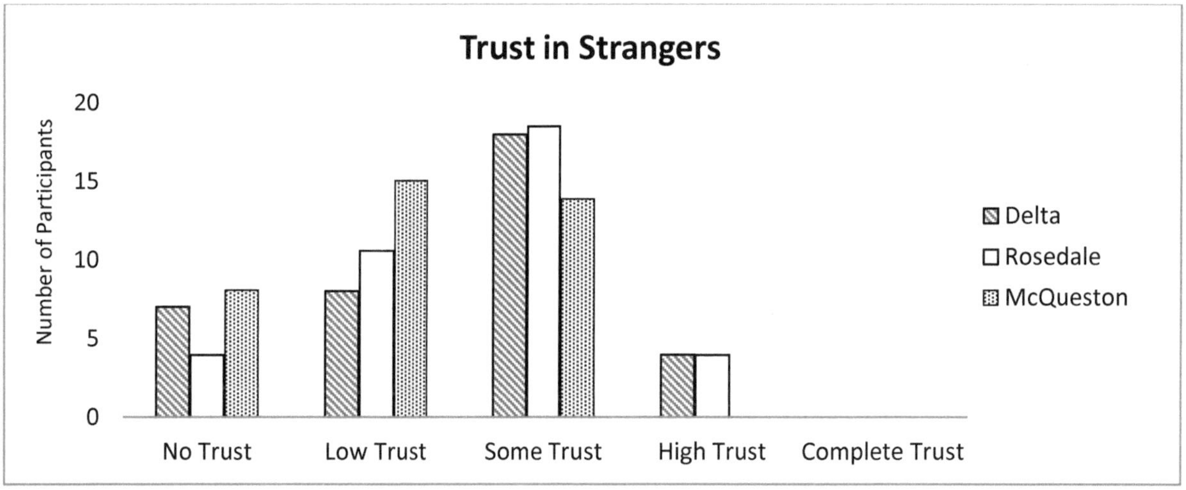

Figure 2-18 - Trust across all relationships and trust in a specific, unknown person or persons across all three Census Tracts (weighted by count). The absence of "Low Trust" and "High Trust" for "Trust People in General" is owing to missing values in the survey instrument. These categories for trust are taken from the Canadian General Social Survey instrument noted earlier.

Participants from McQueston chose no brackets above "Some Trust" with a very few choosing "High Trust" in Delta and Rosedale. The graph reveals a clear bias toward lower trust levels for strangers but the difference between general trust and trust in strangers is not substantial. The following questions about trust are more specific and show greater variation. National responses from 2008 (Cycle 22) a near-even split between a binary of trust people in general (48%) and not trust people in general (52%). The skew toward slightly less trust is consistent with the Census Trusts as well which skew toward not trusting rather than trusting. Strangers, however, engender much lower levels of trust nationally with 42% suggesting that strangers cannot be trusted at all (*General Social Survey Cycle 22: Social Networks (2008)*, 2008).

A different pattern emerges when participants are asked about how much they trust their families and their families, neighbours, and classmates. It is expected that trust in family is greater than trust in strangers. Both categories have some trust, but the skew is strongly to the right of moderate for family and reflects what is held in common knowledge, that there is a strong correlation between relational distance and trust (Figure 2-19).

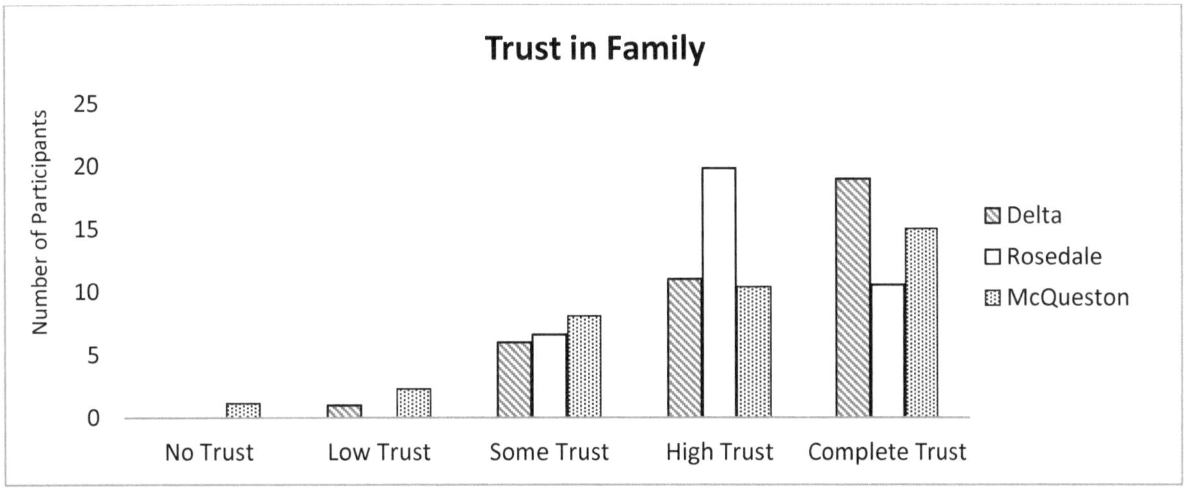

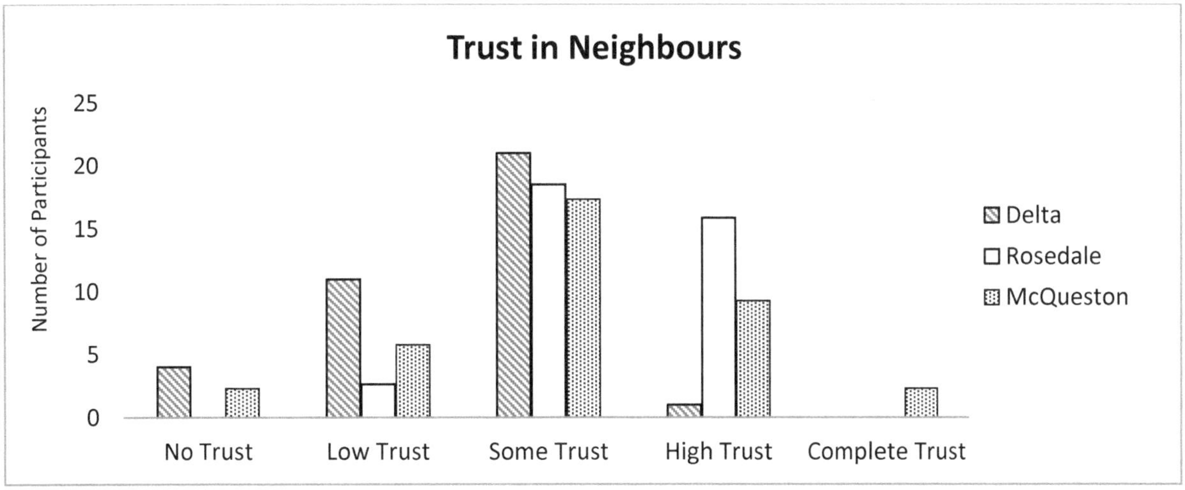

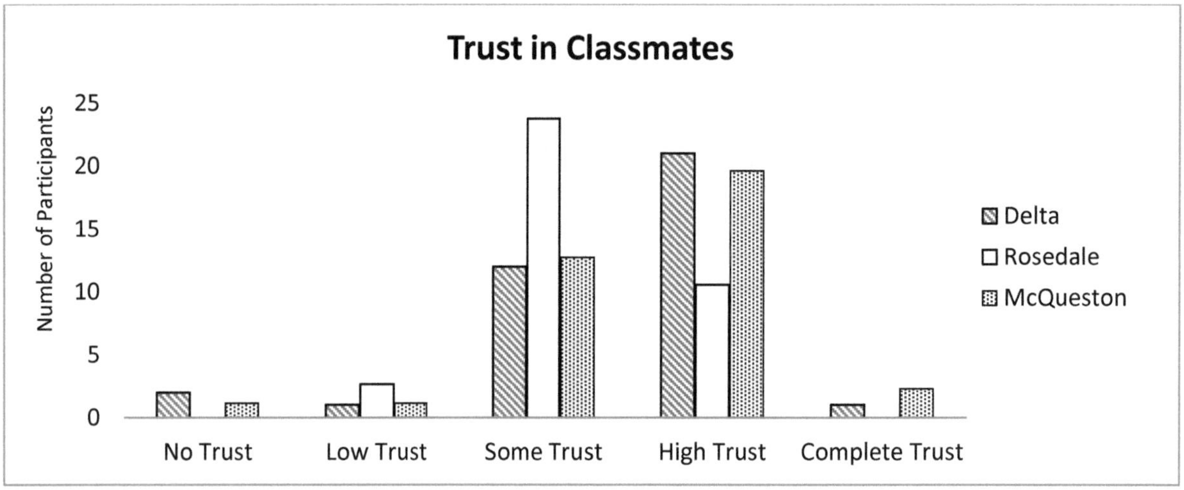

Figure 2-19 - Trust across three distinct community grouping spaces for all three Census Tracts (weighted by count).

Across Canada, the 2008 figures for family trust are 76% for "Can be trusted a lot." Delta's participants showed an unusually high number of "Complete Trust" responses – Rosedale respondents were less willing to draw that conclusion. The category of neighbours tells a different story. Compared with both Rosedale and McQueston, Delta has a significantly lower level of trust in neighbours. Trust in neighbours nationally skews higher than the Census Tracts where 35% commit to a middle level of trust, 46% have greater trust than that and 19% have less trust than that (*General Social Survey Cycle 22: Social Networks (2008)*, 2008). Trust in classmates is a somewhat unusual category and was included given its usage in the GSS cycles. With the average age of participants long past school or college age, this may not be a particularly useful question. The skew toward high trust reflects at the very least a sense that in the intervening years since formal schooling, respondents had a higher degree of trust in their recalled schoolmates than in their neighbours except for Rosedale with a slightly higher level of trust in neighbours.

Institutional trust is a reflection, at least in part, of how the individual perceives the wider society in terms of various organizations that are present in a given local community and with whom the individual has interacted or observed. When asked about their level of trust in the police, the most well represented category for both Delta and McQueston is the "High Trust" bracket (Figure 2-20). Rosedale reflects more ambivalence toward the police with a more moderate trust level. Schools and city officials are similar across all Census Tracts, but schools have a somewhat higher level of trust overall. These core public institutions reflect a somewhat ambivalent degree of trust as perceived by participants with police being the most trusted. Trust in the police is notably higher at a national scale with around 38% of people having a "great deal of confidence" (the highest category) in the police (*General Social Survey Cycle 27: Social Identity (2013)*, 2013).

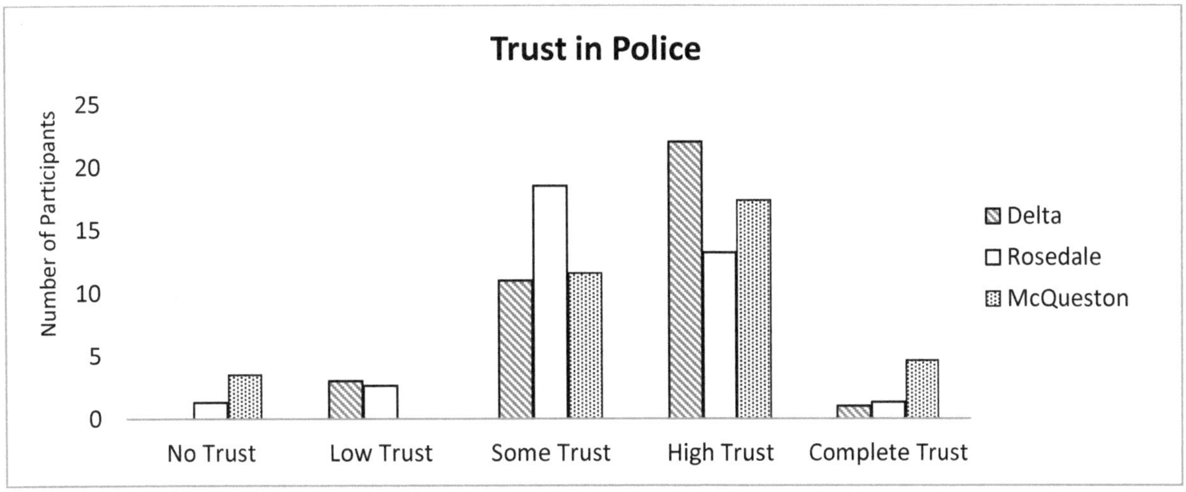

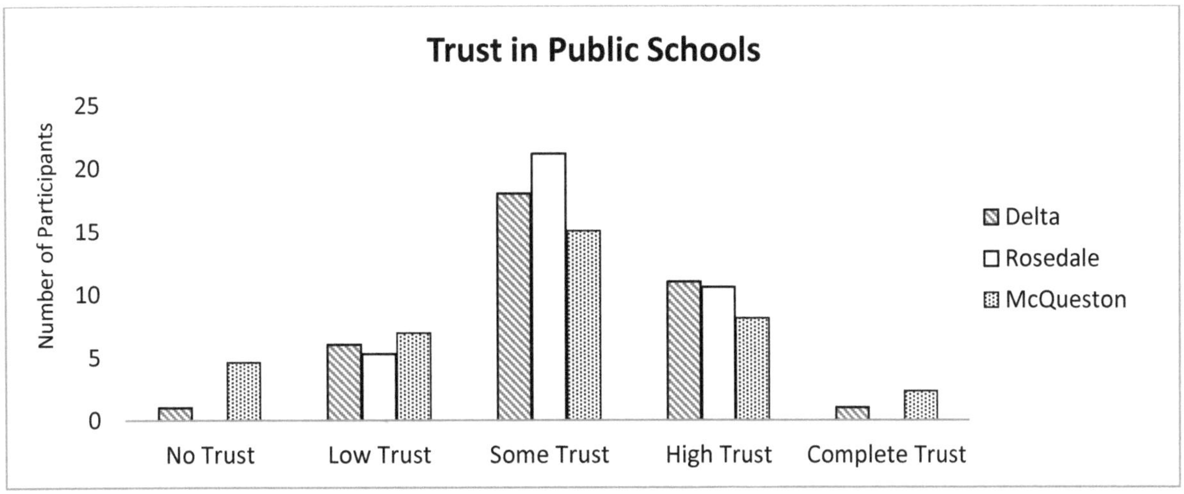

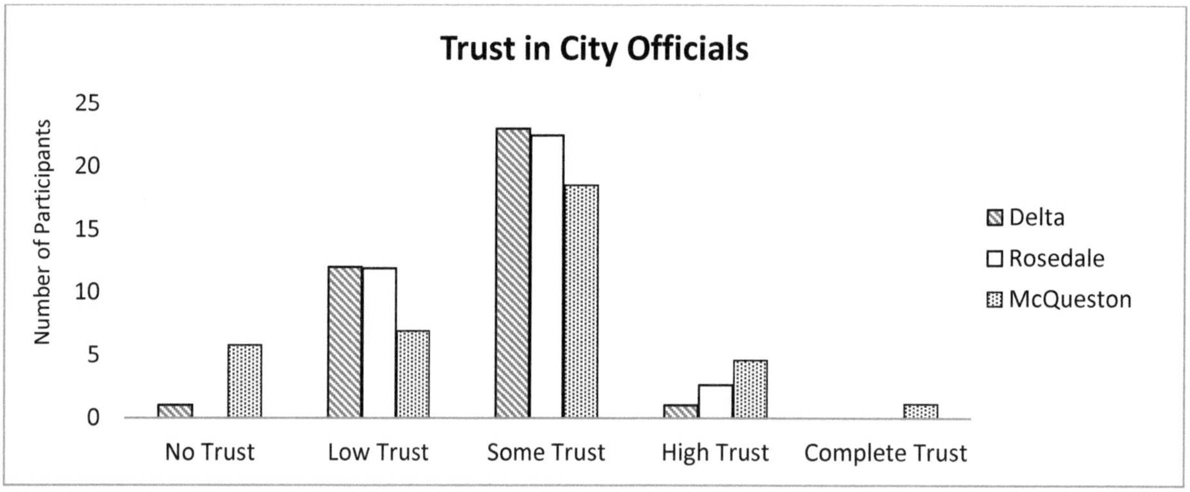

Figure 2-20 - Trust in core institutions - police, schools, and city officials (weighted by count).

If we compare respondent trust levels in the context of local religious groups, local businesses and big businesses, we some interesting patterns. Trust in local businesses shows a remarkably similar pattern to trust levels in schools – moderate but leaning toward high levels of trust (Figure 2-21). Confidence in big business is lower across all three Census Tracts with the higher income Delta neighbourhood reflecting low levels of trust in big business. Trust in local religious groups shows both McQueston and Rosedale as moderate. Delta, by contrast, had a significantly higher level of trust in local religious groups. It is unclear why this would be the case.

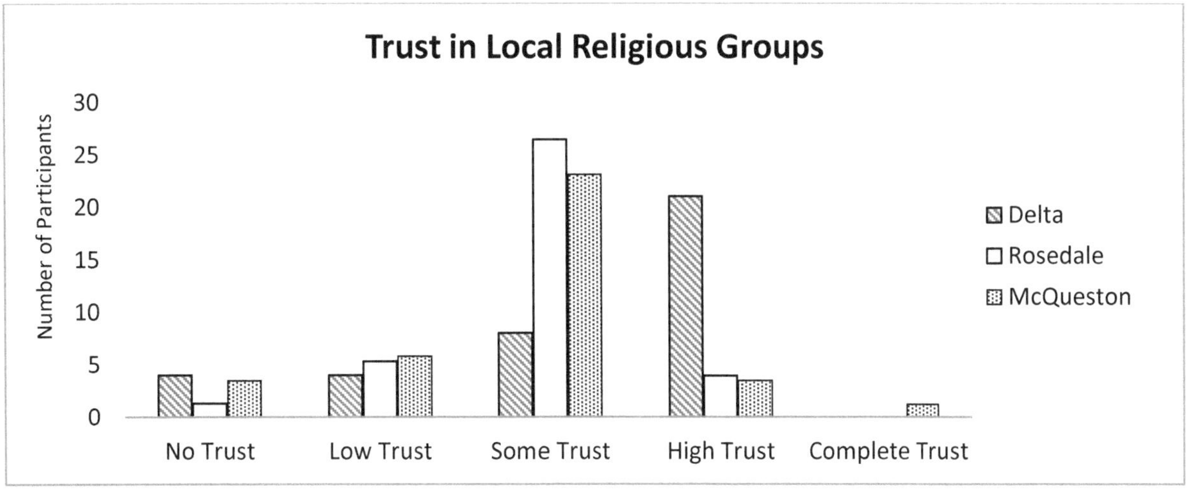

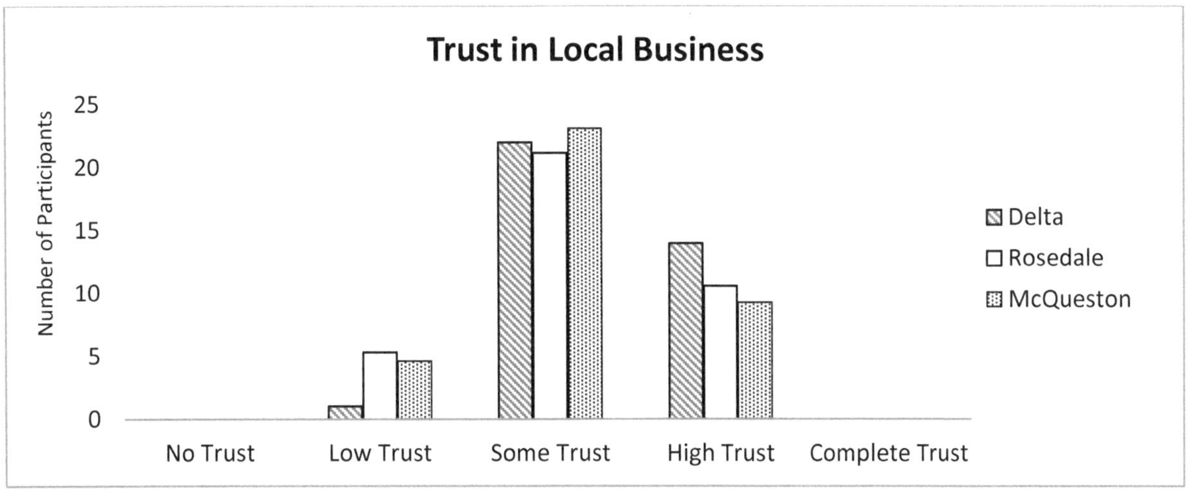

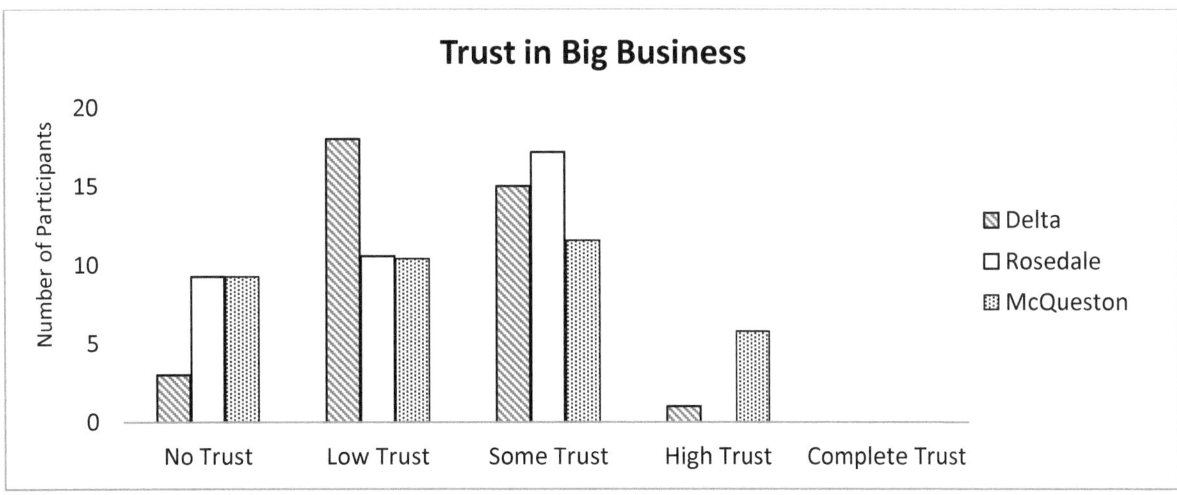

Figure 2-21 - Trust in local religious, local business and big business (weighted by count).

When healthcare and welfare are compared as institutions, participants had common but not equal perceptions of trust across all three. Healthcare is a more trusted institution with Delta showing the highest level of trust. Welfare is a much less trusted institutions, particularly in Rosedale where "Low Trust" was the most represented bracket (Figure *2-22*).

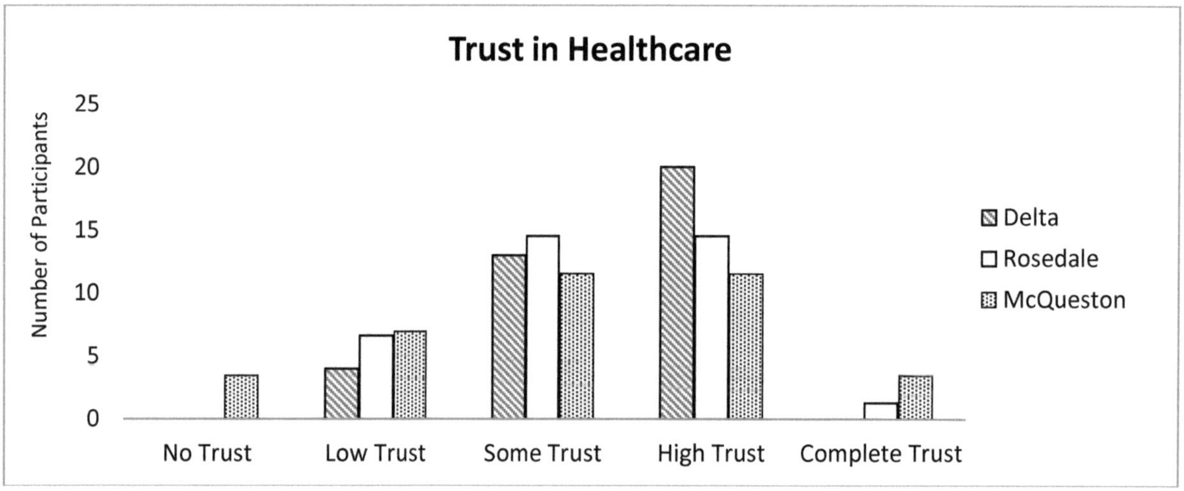

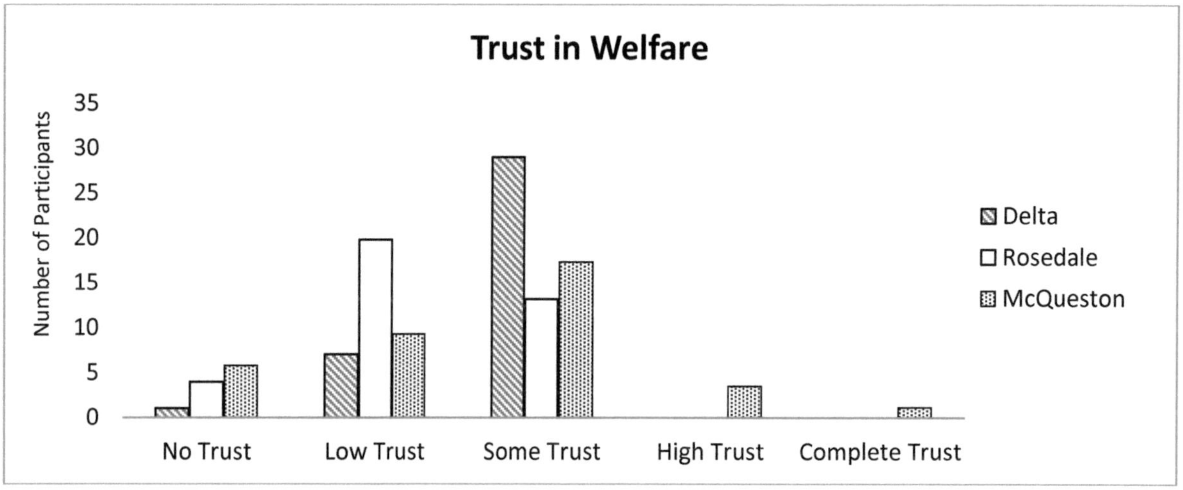

Figure 2-22 - Trust in healthcare and welfare (weighted by count).

2.8.4 Synthesis of Social Capital Measures – Networks and Trust

The foregoing section examined the individual variables for "Demographics and Context" and "Social Networks and Trust." These were compared across the three CTs in the study – McQueston, Rosedale and Delta. A synthetic social capital measurement that aggregates the individual values into a single value was generated to make evaluation of spatial and social capital across CTs clearer. Direct comparison of synthetic social and spatial variables will be made in chapter 4. The synthetic variables are created by combining the natural themes of the social network and trust questions into respective clusters.

Not all social network questions were deemed relevant for a synthetic social network measure. The included questions are Q17, Q21, Q25 because they each reflect participants' evaluation of the size of their social networks which reflects a network **topology (variable: sns.socap)**. The questions that were excluded are Q18, Q22, Q26 because these are categorical questions about type or mode of engagement, a factor that is not relevant to the measure. It was necessary to invert the numeric values of Q19, Q23, and Q27 in order for the increasing magnitude of response to reflect increased relational contact or network **dynamics (variable: ts.socap)** – a participant with no family, friends or acquaintances would have a dynamic score of "0". Physical distance from friends, relatives and acquaintances was also deemed an important social capital measure owing to **proximity** being a significant relational dynamic so Q20, Q24, and Q28 were included – a very low proximity score reflects too little contact according to the question.

Like social networks, the "trust" questions require justification for inclusion or exclusion. Trust in individuals includes Q29, Q30, Q31, Q32, and Q33. The values of Q29 were inverted to reflect higher numbers as higher levels of trust. These questions reflect the perceptions of trust regarding **individuals.** Trust values for **institutions** are also included in the synthetic trust measure. The questions for institutional trust include Q34, Q35, Q36, Q37, Q38, Q39, Q40, and Q41. Where trust in institutions is high, values will move toward 40 as the highest possible level. A value of 8 reflects the lowest possible level of trust in institutions.

Category	Synthetic Variables	Questions	Ranges Low - High	Example 1	Example 2
Social Networks	Topology (**sns.topo**)	Q17 Q21 Q25	0 - 12	7	9
	Dynamics (**sns.dyna**)	*Q19 *Q23 *Q27	0 - 9	4	2
	Proximity (**sns.prox**)	Q20 Q24 Q28	0 - 12	10	4
Sub-Total (**sns.socap**)			0 - 33	21	15
Trust	Individuals (**t.ind**)	*Q29 Q30 Q31 Q32 Q33	5 - 23	19	12
	Institutions (**t.inst**)	Q34 Q35 Q36 Q37 Q38 Q39 Q40 Q41	8 - 40	35	22
Sub-Total (**ts.socap**)				54	34
TOTAL Social Capital Synthetic **(**all.socap**)				75	49

Table 2-8 *Summary table of synthetic capital variables. Variables formatted for use in R. Some values have a minimum of "5" where the lowest value choice on the survey was a 1 rather than a 0. *Original Likert values in the GSS required that these values be changed from ascending to descending before including.**This allows the use of* **all.socap** *as a dependent variable to compare against the remaining independent variables – Q1-16 and the 10 Spatial Variables*

The examples used in the table (Table 2-8) reflect that **Example 1** possesses a synthetic social capital score of 75 while **Example 2** has a lower score of 49. The subsets reflect that **Example 1** has a larger social network and a higher level of trust in both people (19 vs. 12) and institutions (35 vs. 22). These examples show how social network and trust value differences contribute to the variance in the final synthetic social capital scores.

2.9 Results of Synthetic Social Capital

The following section explains the results obtained when social network size and characteristics are integrated and synthesized to produce numeric values that reflect different quantities of social capital. These values enable the provision of an answer to the central hypothesis.

2.9.1 Social Network Size

The results of the synthetic values of the sub-sections reflect variance in the numbers of relationships in social networks by Census Tract. Social network sizes are calculated by taking the middle value of each range and multiplying that by the number of participants in that range (*Figure 2-23*).[7] The ranges arise from the survey instrument where participants are asked to choose the number of:

close friends: 0, 1-2, 3-5, 6-10, 11 or more
acquaintances: 0, 1-2, 3-5, 6-10, 11-20, 21-40, 41-60, 61 or more
relatives: 0, 1-2, 3-5, 6-10, 11-20, 21-40, 41-60, 61 or more

The values of each participant in Delta, Rosedale and McQueston respectively are added together to form an aggregated score for each Census Tract group.

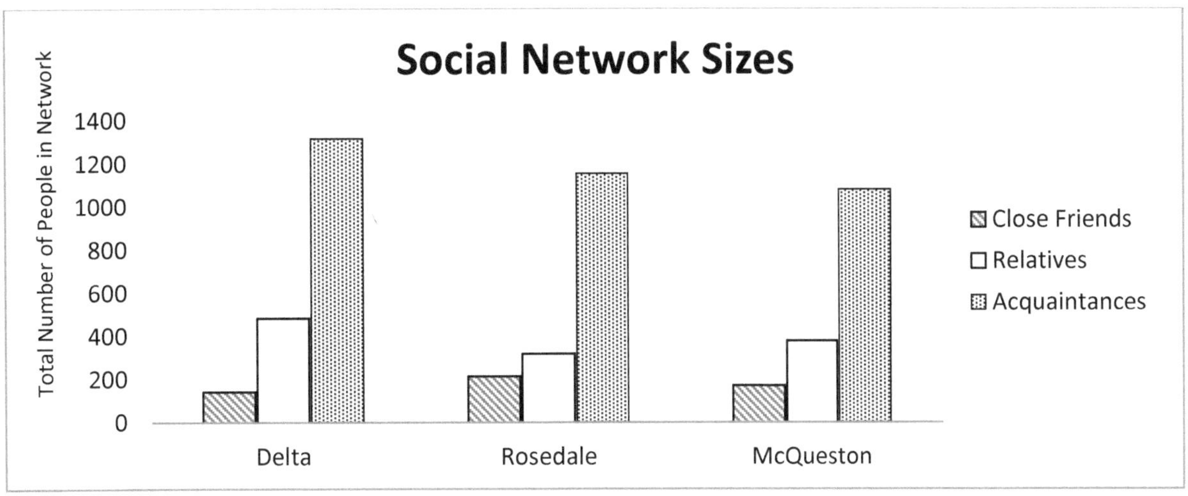

[7] The values are weighted by number of participants but are not normalized for population as the survey numbers, noted earlier, are not sufficient for generalization.

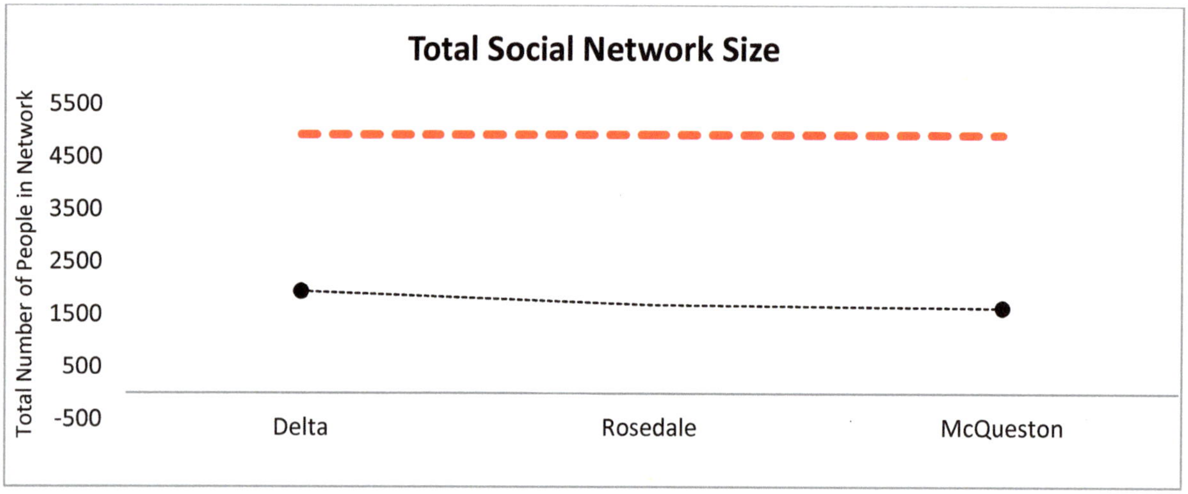

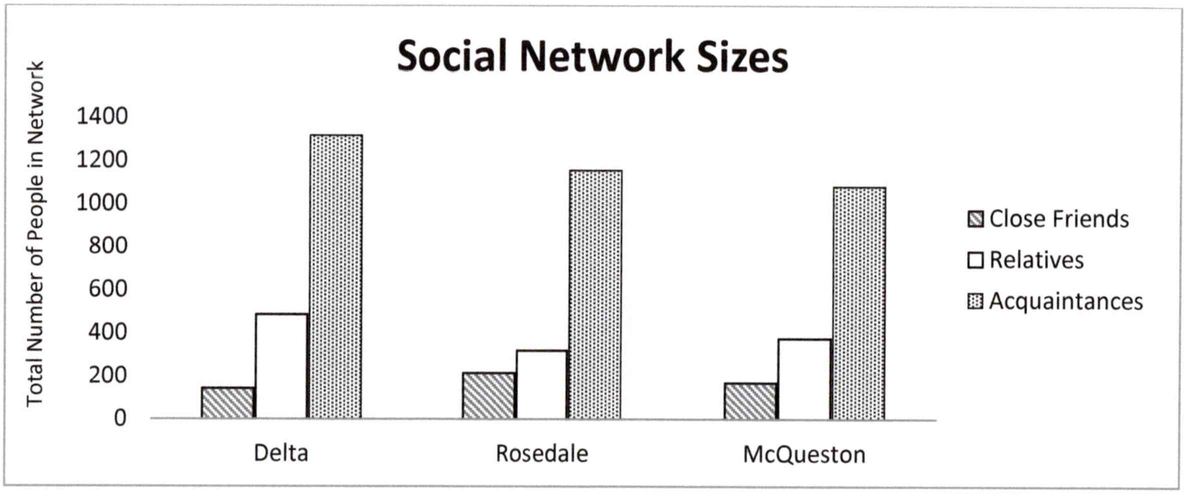

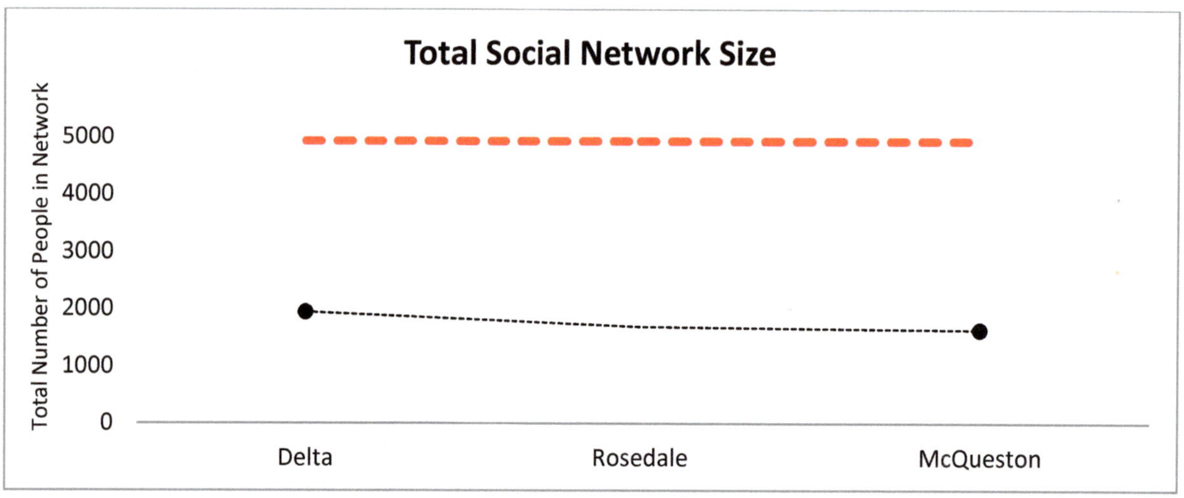

Figure 2-23 - Graph of social network sizes aggregated by respective study areas in both bar graph form with social sections retained (above) and via a trend line where social sections are aggregated.

The maximum potential value (top red line) represents a top limit given the scoring system – i.e. all people chose maximum values. The minimum line would be zero if no social networks were chosen by all participants.

Comparison of social network sizes vary across each of "Close Friends", "Relatives", and "Acquaintances" categories. Total network=cumulative value across all three categories (Table 2-9).

	Rosedale (28)	McQueston (32)	Delta (37)
Close Friends	213	170	142
Relatives	317	376	484
Acquaintances	1155	1079	1316
Social Network Quotient	1685	1625	1942

Table 2-9 Social network sizes (number of people each participant in each of the *close friends*, *relatives*, and *acquaintances* categories) and totals for each of the respective study areas (weighted).

2.9.2 Trust

Trust, a key aspect of social capital, is a critical component of social process at all scales of human life from the interpersonal to the societal. Democracy requires both that individuals voluntarily trust their local neighbours and communities as well the institutions such as police services, banks, educational systems and hospitals. An extensive European study showed that trust can be based on actual events, perceptions of crisis, attitudes toward others, direct experiences or observation of the experiences of others and includes complex interaction between personal and institutional levels of trust (Mingo & Faggiano, 2020). Social behaviours, including how people respond to challenges like climate change, are often dependent on the degree of trust they have in scientists and related experts, a factor that is influenced by those around them, media, and the past performance of people in authority (Cologna & Siegrist, 2020). If trust is high, compliance and positive behaviour change will also improve. The size and inscrutability of significant institutions can create a perception of hidden agendas, manipulated messages and conspiratorial intentions: all of these can undermine institutional trust (Hagen, 2019). Higher levels of trust reflect higher levels of social capital, functioning like a repository for common good, collective actions that can be measured, for instance, by polling attitudes of citizens toward a post-crisis justice movement as seen in a wide Southern American study of the Truth Commission (Mathias et al., 2020).

Recent events tied to the COVID-19 pandemic reflect a concern called "creeping crisis" where a long, gradual accumulation of social stress serves to undermine broad public trust in institutions (Boin et al.,

2020). Loss of trust in institutions can lead to a decline in public safety, loss of coordination in situations that require people to voluntarily endure personal loses for the sake of the common good, and a decreased effectiveness of the institutions in questions – e.g. lower educational system performance, need for increased formal policing, lower compliance with public health initiatives such as immunization where mistrust in existing formal medicine is a stronger driver than belief in the efficacy of alternative forms of medicine (Hornsey et al., 2020). Planners encounter these clashes between expertise and local citizens who may mistrust those experts recommendations on land use changes such as siting wind turbines, high-voltage transmission lines, or land fills (Cain & Nelson, 2013). Given the public service and responsibility domain of planning, interactions between trust (both personal and institutional) are a critical area of concern for professionals carrying out those roles. Consultations with the public reflect wide-ranging attitudes and perceptions ranging from feelings of tokenism to empowerment with many gradients in between have been part of the planning conversation for a long time (Arnstein, 1969). The dynamics that drive these changes are often complex and not adequately understood. For example, a recent study showed that European judicial courts designed to protect constitutional freedoms actually eroded public, institutional trust where they were perceived (whether correctly or not) to be biased or insufficiently objective (Navarrete & Castillo-Ortiz, 2020). Again, whether this is in fact the case is not in question – there need only be a perception of mistrust for negative effects to be realized.

Trust can be very resilient and the exercise of it, perceptive. For example, if a democratic society experiences a violent armed conflict or small scale civil breakdown, it can be the case that citizens lose trust in particular politicians or leaders while retaining trust in the institutions (parliament, police, courts) that comprise their society (Gates & Justesen, 2020). Similarly, when corporations, governments and institutions are perceived as listening to citizens through social media channels and by responding to concerns, trust in those organizations can be increased even where, as in the generation-wide criticism symbolized by Greta Thunberg where a posture of mistrust was the starting point (Reinikainen et al., 2020). Regulation by governments represents a complex balance of personal and institutional trust dynamics. In one example, regulation in the intensified context of Dutch agriculture led to increased trust when local responsibility was devolved to farmers alongside increased quality of interactions between regulators and individual farmers (de Vries et al., 2019). These dynamics require context, however. The relationship between trust, social networks and sharing common goals in one Hong Kong based workplace study discovered that trust played a less important role than the location of a person in their workplace social network and alignment with goals (Chow & Chan, 2008).

Finding ways for city planners to measure social capital more effectively and widely is significant for community trust dynamics. It is vital that planners build approaches to planning that support inclusion of perceptions for both personal and institutional level trust. Wide-ranging numbers of studies that involve social capital strongly suggest that these scales cannot be neatly separated from each other. My study does not comprehensively examine trust or social capital but seeks to improve our ability to describe and quantify social capital at a scale that would be valuable for planners should such a method be found.

The results for "trust" in both the sub-sections and full synthetic scores reflects relatively consistent patterns across the CTs with the highest trust quotient (Delta) being only 1% greater than Rosedale and 3% greater than McQueston (Table 2-10).

	Rosedale(28)	McQueston(32)	Delta(37)
People	595	564	557
Institutions	848	853	907
Trust Quotient	1443	1417	1464

Table 2-10 *Aggregate Trust quotients as totals for people overall and institutions overall. The Trust Quotient is a total of both category aggregates (weighted).*

At the level of trust in people, Rosedale was the highest with Delta 6% lower and McQueston 5% lower. This changes when we look at institutional trust with Delta being 7% higher than Rosedale and 6% higher than McQueston (*Figure 2-24*Figure *2-24* - Trust Quotients compared across Delta, Rosedale and McQueston (weighted by count). Trust quotients calculation: Likert score (1 is low – 5 is high) is multiplied by number of participants in each bracket. Individual trust measures occur earlier in this chapter.).

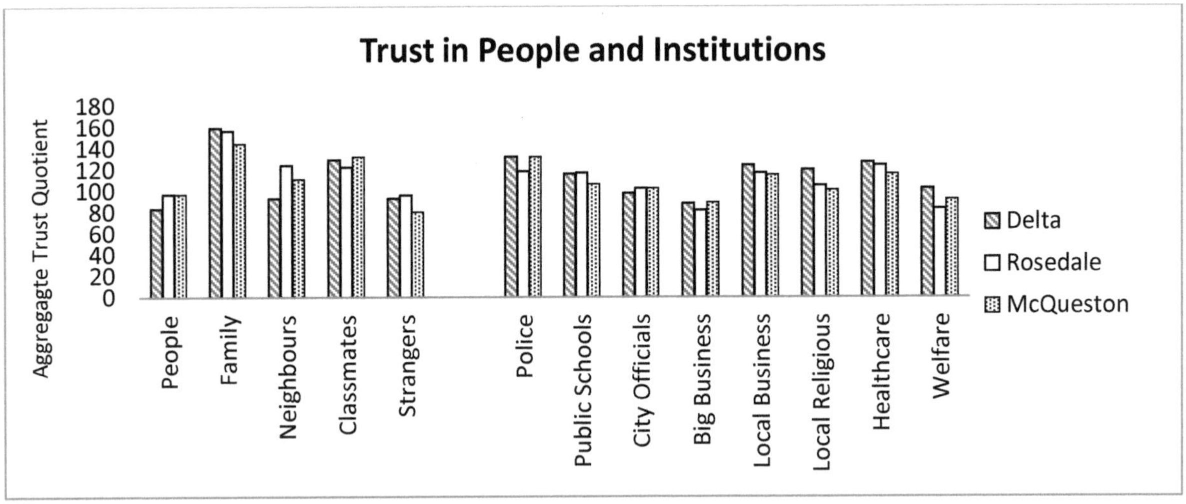

Figure 2-24 - Trust Quotients compared across Delta, Rosedale and McQueston (weighted by count). Trust quotients calculation: Likert score (1 is low – 5 is high) is multiplied by number of participants in each bracket. Individual trust measures occur earlier in this chapter.

From these results, it appears that there is a higher level of trust in people with a median income CT and a higher level of institutional trust in institutions with a higher income CT (*Figure 2-25*).

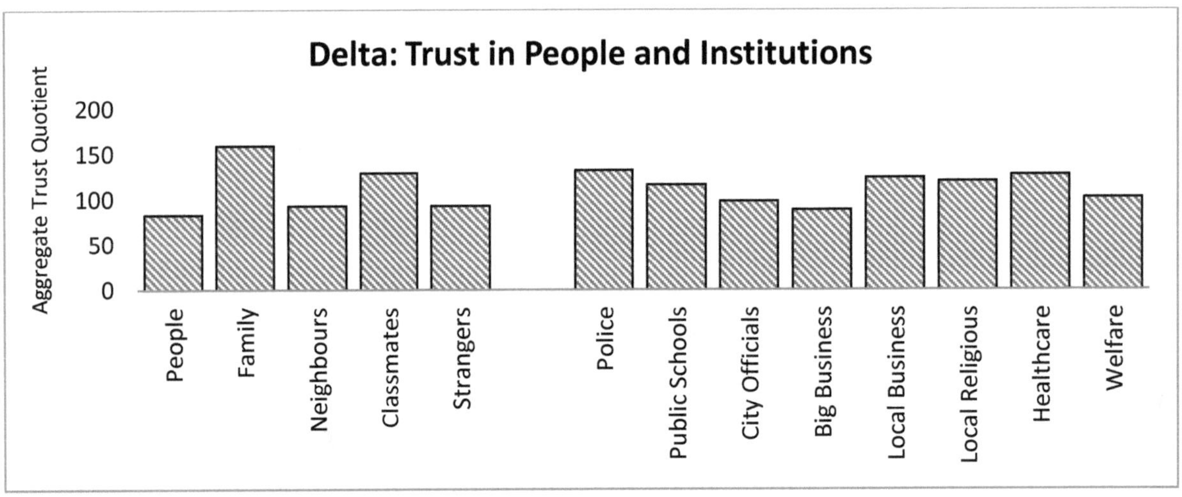

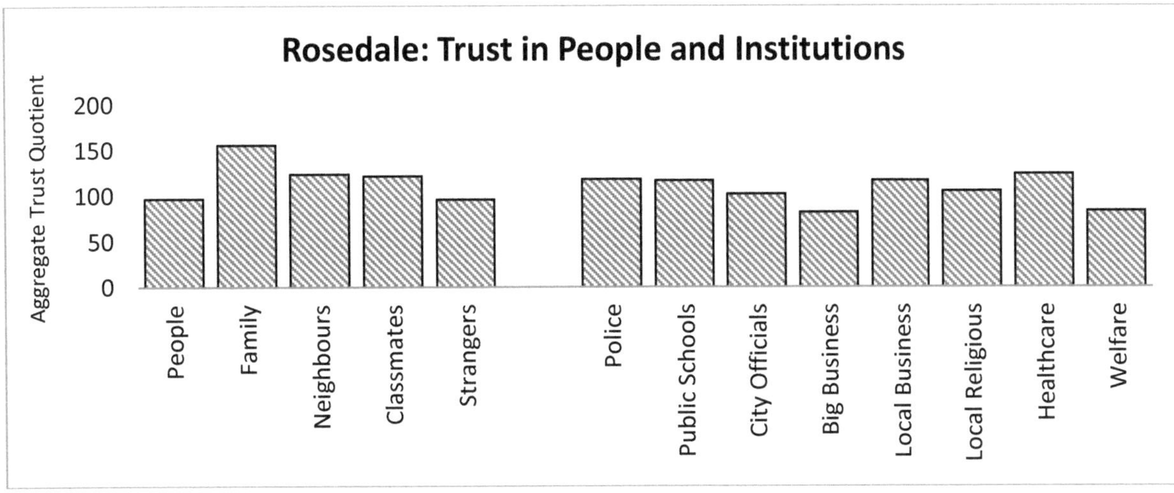

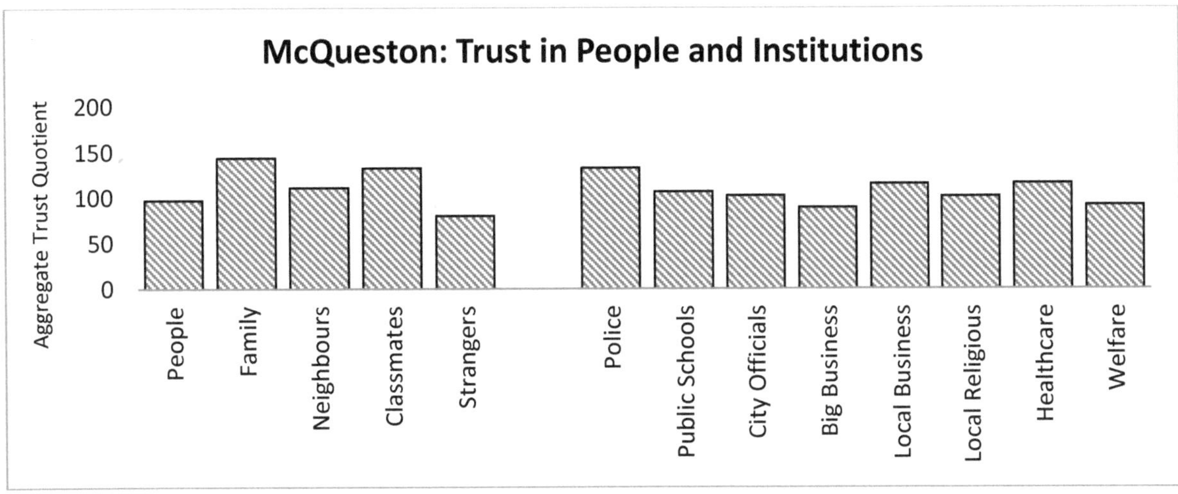

Figure 2-25: Trust Quotients by individual Census Tract (weighted by count) and individual questions combined from earlier individual graphs.

The synthetic measure for trust masks these variances suggesting that where synthetic measures are used, sub-set measures should be included in order to understand possible variations within the measure (Figure *2-26*).

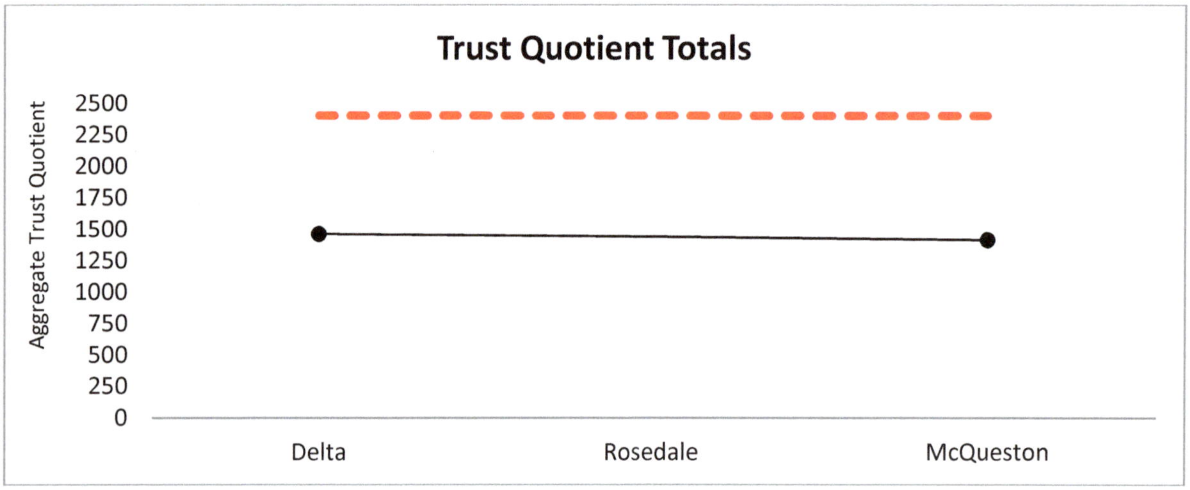

Figure 2-26 - Aggregate of all Trust scores. Red dashed upper line represents maximum amount of trust where all responses are total trust (5) on the Likert scale – an upper limit based on the survey value options.

2.10 Discussion

The objective of the research design was to support verification of the central hypothesis, that social capital levels are higher where spatial movement is greater and the secondary hypothesis that income and social capital do not rise and follow together in a linear way. In this later case, there is a uniformity of social capital across all three Census Tracts. This supports the secondary hypothesis weakly because social capital does not decline appreciably from the 75th to 25th percentile income bracket. The scale I focus on is neighbourhood / Census Tract aggregate scoring so that we may find below that there are signals of individual social capital differences that are correlated with income. In particular, Delta (high) has a higher overall social network size and as noted in earlier sections, trust varies at a more granular level. One key conclusion from this analysis is that aggregated social capital scores are of lesser value than scores on specific themes such as trust or social network size. Variation appears to get lost when averaged across multiple participants. Higher levels of trust in the police and local religious institutions characterize Delta (high) while social network sizes change little across close friends, relatives, and acquaintance networks regardless of Census Tract income levels. Statistical explorations of spatial and social data may reveal interacting variables in individual cases that are lost when Census Tracts are treated as aggregates.

Variations are more notable when survey results are compared across Census Tracts on a question-by-question basis. Though they share a common urban form and geographic proximity, age, income, levels of trust, social network characteristics, and sense of belonging vary across the three CTs. Delta has a significant income bracket for the 60,000-79,999 range, McQueston (low) in the 15,000-29,999 range, and Rosedale (mid) has a much more distributed representation across the 5,000-99,999 range. Delta (high) has far fewer people who say they have neighbours they could ask for a favour, Rosedale (mid) has more in the 1-5 neighbour category and McQueston (low) has slightly more than 5. Despite this variation, a sense of belonging is nearly even across the three CTs. Relationally, having too much contact with any group is rare and McQueston (low) has double the number of people who say close friends mode of contact is face to face. Finally, it would appear that acquaintances are much more likely to be within walking distance than close friends or relatives, suggesting that close friend and relative relationships persist despite geography while acquaintances are much more oriented to geographic proximity. This can be an important and positive dynamic for communities, particularly where acceptance of diversity is concerned (Peterson, 2016). The high number of face-to-face contacts for acquaintances bears this out. As we continue to explore social capital, it may well be that the ability to distinguish types and modes of social capital dynamics, perhaps even a more fully developed taxonomy of social capital, will be as critical as finding broad pattern proxy measures.

Chapter 3: Ten Exploratory Metrics for Comparing GPS Spatial Use Patterns with Socio Economic Data in Urban Settings

3.1 Introduction

The purpose of this chapter is to apply a range of spatial metrics to the week-long GPS tracks generated by study participants. The central hypothesis, that *social capital levels decrease with decreasing spatial movements*, requires that both a social capital measurement (provided in Chapter 2) and a spatial measurement are acquired. In this chapter, spatial measures that can be compared with the social capital data will be secured by characterizing the time and location data of participants in ten different ways. These numeric summaries are in turn compared with the demographic portion of the social capital survey such as age, years in the neighbourhood, income and gender. The three distinct areas (Delta, Rosedale, McQueston) will be examined uniquely. Direct comparison of social capital (social networks and trust) and spatial patterns is the subject of Chapter 5.

3.2 Methods for Spatial observation through GPS data logging

One of the benefits of logged GPS data is a behavioural record of physical movement over a geographical area in a given time period. This empirical data provides an objective record of human motion that can be analyzed and compared with other demographic, context and social data. There has been a growing focus on the qualitative and quantitative aspects of GIS data as a rich intersection for theory and practice (Pavlovskaya, 2006). This chapter contributes to the exploration of this intersection through collection of an original data set that is both quantitative and qualitative. Although apps and social media data tracking is possible, the objective of a random sample at a time when such devices were not ubiquitous (difference of app use by age, ethnicity, gender) meant that supplying a passive common device was a more disciplined way of obtaining the data. Emerging positional data collection through the Internet of Things, mobile phones, and other forms of passive location sensing in urban environments (Erturk et al., 2019; Lavric, 2019) provide new options for furthering the central and secondary hypothesis that were not available at the time of this study.

In order to be included in the study, recruited participants completed the Social Capital General Social Survey (SCGSS) instrument (Friesen, 2018b) and then carried a GPS data logging device with them for 7 days logging at 15 second intervals. A one-week collection period was used owing to common cultural practices oriented around a weeklong cyclical time-scale pattern. This framework may not reflect actual small-scale time patterns for all participants and other insights may be gained by using longer or shorter time periods. Within the limits of this study a 7-day period represents a reasonable working assumption and follows GSS Time Use Survey practices (e.g. Canadian General Social Survey 2010 – Time Use Survey, n=15,390) and other spatial data approaches (DeBats, 2011; Kalkhan, 2011; Kwakkel et al., 2014; Wedding et al., 2018).

The GPS data logging instrument was selected after a process of designing and building a prototype designed to deepen awareness of the technical function of GPS devices. The prototype hardware and software was based on an open source development platform. This platform has significant community development support, makes use of the C++ physical computing language, and is continuously expanding through a growing network of developers, engineers, electronics retailers, researchers and educators (Margolis, 2012; T. Nguyen et al., 2018; Rosa et al., 2017). Professor Darren Scott, Faculty of Science, School of Geography and Earth Science (McMaster University) generously offered the use of TransLAB's 91 QStarz BT-Q1000X Travel Recorders for the data collection. The BT-Q1000X devices use a MTK II GPS module with wide channel range, high sensitivity antenna functions and internal memory capable of logging ~200,000 GPS data points, including adjustable sampling intervals which were set at 60 seconds. The device required nightly charging and participants were instructed to take the device with them whenever they left the house. No tracking was possible with the device. Data could only be logged passively to the device and data downloaded after participants returned it to the collection firm. They were asked to do so after they had been using the device for at least seven days.

Participants also filled in a paper time-use diary (included in GPS device mail-out) as a means of corroborating the GPS observational data with contextual information such as primary activity and the extent to which participants identified their data collection time-frame as a normal week (D. M. Scott et al., 2013). Time-use dairies are a long-standing method for enriching study participant details across health, life-span and occupational ranges (Chilvers et al., 2010; Robinson, 1999; M. Thomas et al., 2011). In this study, the diary corroboration was designed to identify anomalies such as unusual weather, construction, illness, or other unusual life circumstances for each participant as well as for the study area in which they lived. Studies indicate that if diary use is too detailed, participants tend to be less accurate (M. Thompson & Oelker, 2013). Although researcher-directed follow-up interviews have been shown to improve recall on daily pattern details (Orban et al., 2012), in this study, the purpose of the time-use diary was simply to provide corroboration that the patterns of participants were typical, not subject to some unforeseen disruption that would have impacted potential results across the various study areas. Time use at an individual level beyond these general patterns is not within the scope of the research.

Out of the 97 participants, 50 did not provide any additional comments beyond noting that their week had been *normal*. There were no comments that suggested notable disruptions to local travel patterns during the data collection period. The only variance was a slight skew toward *less active* but given the variance in typical respondent driven perception data, the significant result was the *normal* finding (Figure 3-1).

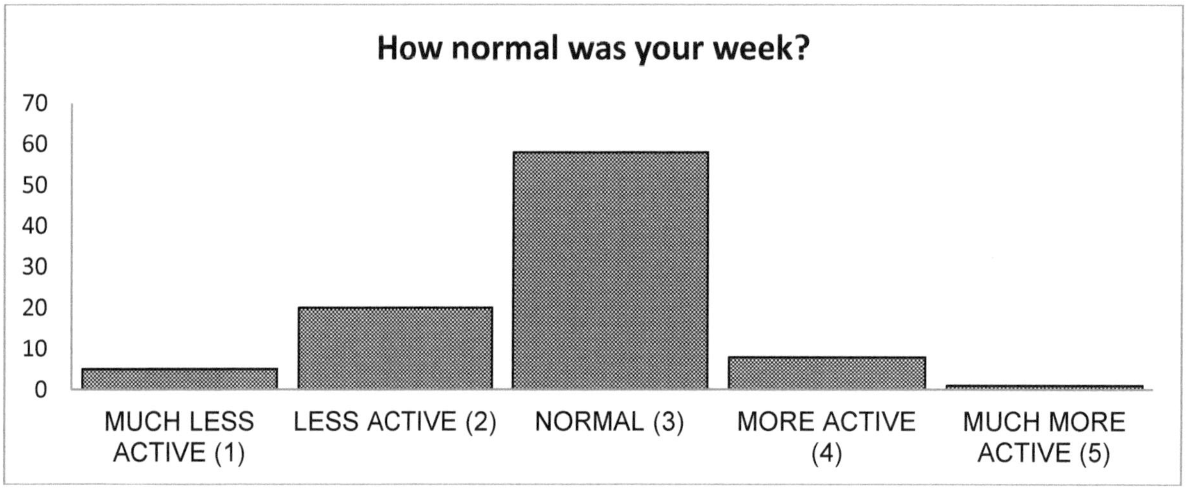

Figure 3-1 - Participant responses to the normalcy of the week in which they carried the GPS data logging devices (all participants).

3.3 Time Use

The diary design employs four dominant time use categories identified in time use surveys. categories that are the most dominant in other specific time use studies (Bechard & Marchand, 2006). These categories are *work*, *home*, *school* and *other* as an open write in to capture the wide range of low frequency, high variance possibilities that this variable represents (*General Social Survey - 2010 Overview of the Time Use of Canadians*, 2011). The time blocks are listed M (morning), A(afternoon), E(evening), and N(night) with a numeric value identifying the primary activity: (1) is *work*, (2) is *home*, (3) is *school*, and (4) is *other*.

An example of the time use diary (noted by time at location) data for one participant revealed that there was no time spent in school but there were 16 blocks of six hours each there the primary activity was home. The patterns of time use support what is seen in the spatial data. When examined as a total group, the 97 participants spent most of their time at home with a balance of work and other activities and very little school.

When the three Census Tracts are compared on the basis of time use primary activities, they reflect a significant common pattern with Delta(High) showing slightly more *work* time blocks, McQueston(Low) showing slightly more *home* and Rosedale(Mid) slightly lower on *home* and *other* (Figure 3-2).

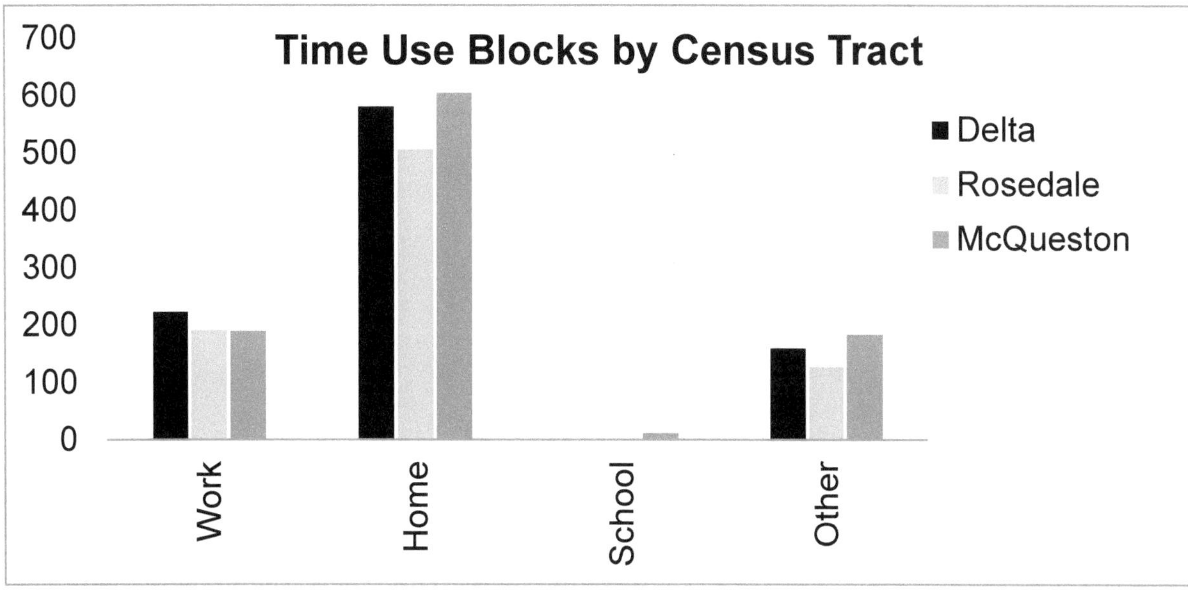

Figure 3-2 - Comparison of three study areas in terms of time use blocks (weighted by number of participants in each Census Tract) - Delta(High), Rosedale(Mid), McQueston(Low). The x-axis represents the number of 6 hour blocks designated for each primary activity.

3.4 Data Collection

Research participants were recruited through a random digit dialing (RDD) telephone strategy. Given the need to establish clear geographical location within respective Census Tracts, the population was restricted to residents with landlines. A market research firm conducted the telephone surveys, mailed GPS devices to participants, and then removed data from the devices when they were returned. This survey and GPS data was compiled along with a summary of call records and delivered for analysis.

Collection of data from the field began on January 20, 2016 and concluded on March 23, 2016 with late device submissions ending May 4, 2016. With a limited number of devices, a rolling schedule was followed that saw returned devices cleaned and sent to new participants as they became available. The graph of collection periods reflects values that have been adjusted to the seven-day period and their relative distributions over the three months of data collection which was required (Figure *3-3*).

In some cases, participants carried the devices far longer than instructed and in others less. The data set was reviewed and participants with 1-2 days of data were excluded unless diary comments gave a clear indication for the gap in collection, e.g., being home as a normal habit so that the 2 days of data were part of a typical 7-day week movement pattern. Any participants with brief collection periods who did not note reasons for the low movement were excluded from the study. There may be movement changes seasonally across an annual cycle but January to March as the core collection period reflects a common seasonal bracket for participants. Variances in the "Other" category were not explored in more detail given that the focus of the study was not the specific details of a given participant but rather the aggregated collection of travel paths within the study areas as noted earlier.

The average number of data collection days for GPS data was 6.7 and the mean average number of observations (GPS logging points) after standardization was 21,657. The median average was 22,194 per participant. The total number of GPS data logged observations across all participants after these adjustments was two million one hundred thousand.

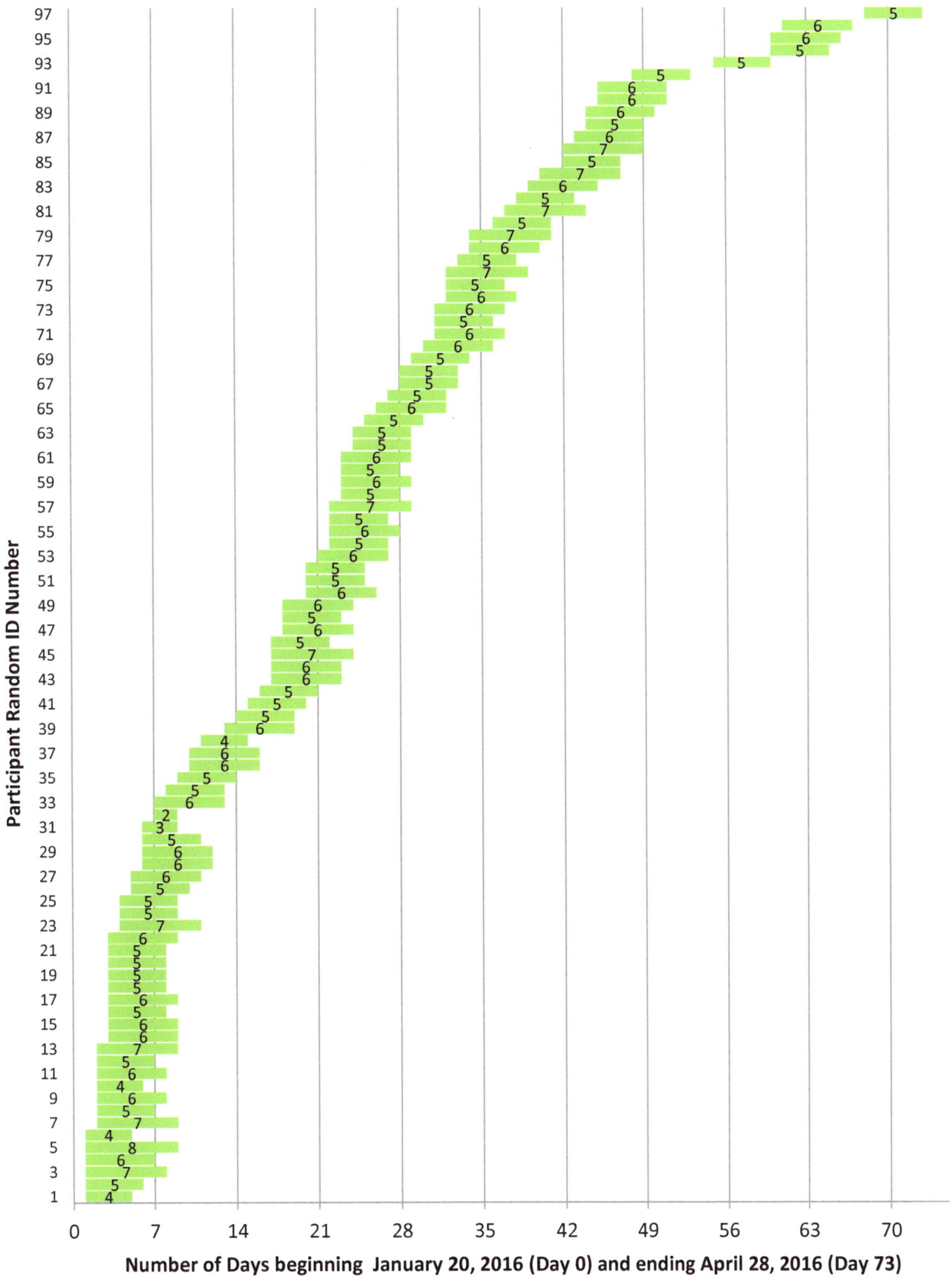

Figure 3-3 - Time based distribution of data collection periods for each of the 97 participants (y axis = privacy ID for each participant). For example, the lower left participant AT25N began GPS data logging on January 20, 2016 and ended (4) days later on January 25.

3.5 Spatial Data Processing

With the Census Tract characteristics and population profiles established, our examination will now turn to the spatial characteristics of the participants. There were two primary steps taken in the spatial data cleaning process (G. Christensen & Miguel, 2018; R. Christensen et al., 2010; Hastie et al., 2011; Patil et al., 2016; Wedding et al., 2018; Young et al., 1978). The first involved reviewing the data from each of the 97 participants to align with the 7 days of recording at 15 second intervals. Each file was examined and processed to align as closely as possible with this requirement. Some participants turned devices on only when travelling while most left the devices running even while home, an accepted practice explained in the contact telephone call and explanation letter that accompanied the mailing of the device. The second data processing step was to convert the CSV files generated by the device into a format suitable for ArcGIS analysis. The raw data began with twenty-three million two hundred thousand observations of 11 variables (Table 3-1).

	A	B	C	D	E	F	G	H	I	J	K
1	INDEX	UTC DATE	UTC TIME	LOCAL DATE	LOCAL TIME	LATITUDE	N/S	LONGITUDE	E/W	ALTITUDE	SPEED
2	1	20/01/2016	12:26:39	20/01/2016	7:26:39	43.237284	N	79.818329	W	92.328178	0.87415
3	2	20/01/2016	12:26:44	20/01/2016	7:26:44	43.237234	N	79.818397	W	122.671326	0.468212
4	3	20/01/2016	12:26:49	20/01/2016	7:26:49	43.237211	N	79.81834	W	122.574478	0.61825
5	4	20/01/2016	12:26:54	20/01/2016	7:26:54	43.237195	N	79.81832	W	121.10376	0.674647
6	5	20/01/2016	12:26:59	20/01/2016	7:26:59	43.237156	N	79.818294	W	119.720085	1.335589

Table 3-1 *Raw data format as downloaded from BT-Q1000 X devices.*

A custom Excel macro was written and tested (Appendix B) to handle the processing of 97 individual files, some of which had more than 40,000 observations. The primary function of the macro was to remove unwanted sub-variables (UTC Time, cardinal columns, altitude) and then format the desired sub-variables (index, date, time, latitude/longitude decimal places) for ArcGIS import. After processing with the macro, the final data set was reduced to two million one hundred thousand observations of 5 variables across all 97 participants.

Spatial statistical tools were developed and tested in ArcGIS to process the 97 data files and generate output features, tables, images and reports. Reports and tables provided the detailed summaries of spatial statistics for each participant track and were included in the data frame holding demographic, context, social network and trust data from the survey. Data files were organized into three sections – Delta (High), Rosedale (Mid) and McQueston (Low) – and that structure was used for organizing and sorting the data processing that followed. The data frame setting used for the map was NAD_1983_CORS96_UTM_Zone_17N.

ArcGIS tools apply transformations to the CSV data files of each participant to produce maps and analytic summaries of the spatial data. Calculation of measures such as average speed were made using the "Summary Statistics" tools. Each of the three Census Tract files were imported into a geodatabase to facilitate iterative processing of individual participant files into core spatial data outputs which suitable for comparative analysis. These outputs included values that were used in custom Excel formulas to establish unique new variables such as *Distance between Median and Mean*, and *Difference between X and Y*, which are not measurement tools available in ArcGIS. The details of these spatial variables are explained in the following section.

3.5.1.1 Rationale for Spatial Analysis

One of the challenges of comparing social data and spatial patterns is determining where the point of comparison should be (Knigge & Cope, 2006; Pavlovskaya, 2006). As noted earlier, commensurability is an important field of critical study for sociology and is concerned with examining how two different dynamics or characteristics may or may not be compared by means of a common numeric value (Wendy N. Espeland & Stevens, 1998). Consideration needs to be given to the difference at two levels. The first level is represented by individual tracks which can be described and analyzed individually (E. H. Chung & Shalaby, 2005). This analysis highlights the characteristics of the spatial track that are amenable to statistical analysis as spatial rather than just numeric distributions (Kwakkel et al., 2014).

A second order of analysis is examination of clusters of tracks. In this case, the clusters are comprised of survey participants from respective study areas who are treated as a group where the clustered tracks are analyzed in an aggregated form. Analysis of the ten variables across the 97 participants, and then across the three Census Tract, clusters supports formal description of the spatial patterns in ways that can then be usefully compared with the aggregated SCGSS survey data (Axhausen, 2008; Blokland-Potters & Savage, 2008). Time use dairy entries, noted above, are intended to verify a normal week in terms of activity levels rather than tracking personal activity details or the specific locations of these activities.

Time and place are vital to social capital development and ongoing dynamics(Dugundji et al., 2012; Farber et al., 2014). This was examined in detail in the Introduction 1.1 and following. Trust and social networks of humans interacting with other humans are deeply spatial and are therefore important dynamics to understand as new tools and methods for capturing that data are developed(Dugundji et al., 2012; Harrison et al., 2019; Logan, 2012). New analytic techniques which are functional with large datasets characteristic of space and place studies means we are in a fruitful era for social capital study(Adrienko & Adrienko, 2011a). This is reflected in the large and rapidly growing social capital literature. Neighbourhood scale focus depends on individual participants as contributors to that neighbourhood (Census Tract) scale. Translating social capital dynamics into analyzable numeric forms and capturing a diverse range of mathematical features for the spatial path networks locates the measure at that suitable neighbourhood scale without exhausting insight that would be available at single participant levels had that been the focus of the study – where they work, what they do outside of their home when not working, what their shopping or socializing habits are outside the home, and so on (Axhausen, 2008; Omrani, 2015; D. M. Scott et al., 2013).

Likewise, the central hypothesis is concerned with comparing aggregate spatial measures and social capital, regardless of details like purpose or specific activity. This focus on finding ways that various summaries of spatial pattern characteristics that correlate with spatial capital is a key aspect of the experimental nature of this study. The range of ten unique spatial features increases the possibility of discovering a useful relationship(Bennett & Vale, 2015; Ester et al., 1996; March & Steadman, 1974; Miller, 2004). These are coarse metrics compared with individual paths and detailed time-use or specific social interactions but allow for pattern evaluation at a neighbourhood scale. Details in the study are such as most-used pathways or barriers are too fine-grained to be considered. Social interaction spaces are also excluded since general social capital patterns are examined rather than considering how coincident time and space interaction probabilities with given participants may change from place to place (Crandall et al., 2010; D. M. Scott et al., 2013; Nj Thrift, 1983).

3.6 Spatial Variables Selected for Analysis

There are ten numeric outputs of from the models used in analyzing the spatial patterns created by GPS data logged participant tracks (Table 3-2). These ten measurements arise from two primary sources: 1) ArcGIS Network Analysis toolkit; 2) Synthetic variables based on the ArcGIS tools. For a measure to be included, it needed to represent a measurement that examined a distinct dimension of the complex spatial dynamics of human movement in an urban setting while being as simple and clear as possible (Stanley, 2007) and which was reducible to a single numeric value that could be included regression, logistic, random forest, and other models.

Variable (variable.name)	Explanation	ArcGIS Tool	Example Outputs
Total Distance (total.distance)	Length of line generated by GPS tracks	Points to Line	71509 meters
Radius of the Standard Distance from the Mean (standis.radius)	The distance from the mean geographic center of a set of coordinates to the outer edge of a circle with radius equal to one standard deviation from the mean	Standard Distance (Figure 11)	469 meters
*Difference between the Mean and the Median (dif.memed)	Distance between the location of the geographic mean and the median set of coordinates.	(Calculated with custom Excel formula)	63 meters
Rotation of the Standard Distance Ellipse (ellips.rotat)	The rotation clockwise from North (0 degrees) of the standard deviation of the x axis when calculated as a value of the standard deviation of the y axis	Directional Distribution (Figure 12)	80.6 degrees
*Ratio of the X and Y axis of the Standard Deviational Ellipse (ellips.xyrat)	Length of standard deviation of the X axis divided by the standard deviation of the Y axis where a perfect circle would equal 1.	(Calculated with custom Excel formula)	7.4 (longer x axis vs. y axis)
Nearest Neighbour Observed Mean Distance (nn.obmedist)	Computes the average distance from one point to all others for the whole set and then identifies the median distance.	Average Nearest Neighbour (Figure 13)	0.540 meters
Nearest Neighbour Ratio (nn.ratio)	Compares how close the next nearest point is compared with a random distribution of points. Clustered < 1 < Dispersed	Average Nearest Neighbour	0.050
Nearest Neighbour Z-score (nn.zscore)	Measures how likely it is that a given distribution of points is random. Low scores equal very unlikely distribution is random.	Average Nearest Neighbour	-278.89
Nearest Neighbour Area (allpointsarea.rectangle)	Area of the smallest rectangle that encloses all points – does not necessarily align with x and y axes.	Average Nearest Neighbour	10960790 square meters
Standard Speed (standard.speed)	Calculations of all the speed values carried into the dataset from the GPS devices expressed as standard deviation.	Summary Statistics	3.52 km/h

Table 3-2 Variables used for spatial analysis – data table (variable names) as used in R-based coding of models and data frames. Detailed descriptions of variables are available (Ebdon, 1991; Mitchell, 1999, 2005; L. Scott & Janikas, 2010). *Synthetic variables developed for this paper.

3.6.1.1 *Total Distance*

The *total distance* variable is calculated by adding up all the distances between GPS features for every participant. The calculations were made using the "Points to Line" feature in ArcGIS 10.3.1 and are based on Euclidean measurements (Equation 1) after features are converted to XY points and projected as noted above.

Equation 1:

$$distance = \sqrt{(x_2 - x_1)^2 + (y_2 - y_1)^2}$$

The total distance measurement established how far a given participant travelled outside of their home over the course of the data collection period. There are potential interactions with this variable and other dynamics such as mode of travel and type of work engaged in. The measure, however, is of primary value in determining range and intensity of travel habits in a single, linear measurement.

3.6.1.2 *Standard Distance from Mean*

For a set of geographic coordinates it is possible to calculate the mean center of a set by comparing the distances between each point and the mean center, and then averaging those distances (Figure 3-4) (Bennett & Vale, 2015; Enedy, 1993).

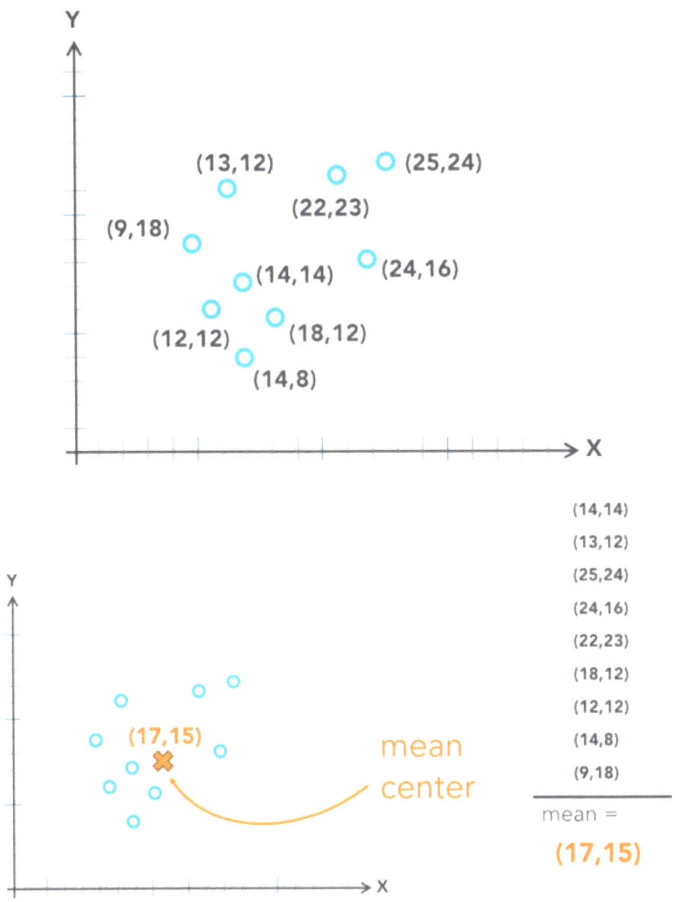

Figure 3-4: Calculating the Mean Center for a set of geographic coordinates. Figures are modifications from ESRI technical workshop graphics (Bennett and Vale, 2015).

A standard deviation from that mean can then be determined and is known as the *standard distance from the mean* (Figure 3-5).

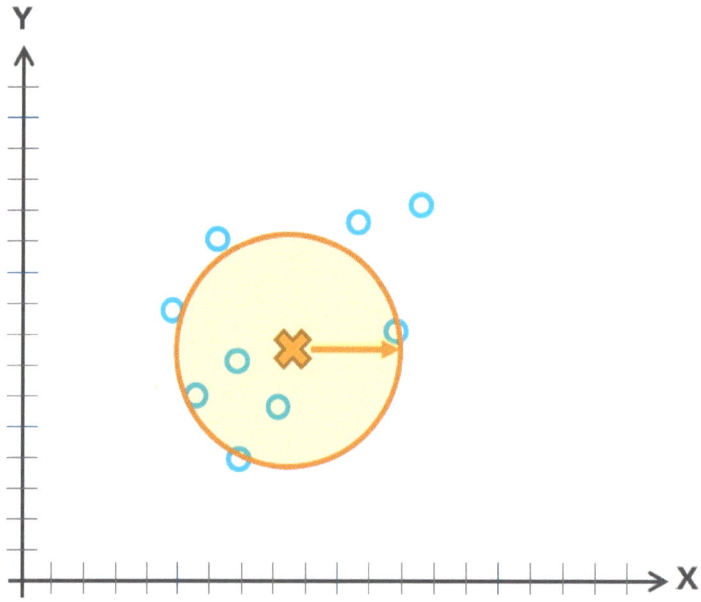

Figure 3-5 - Standard distance from the mean - radius of circle around mean centre of GPS features represents one standard deviation. Figures are modifications from ESRI technical workshop graphics (Bennett and Vale, 2015).

This calculation gives each participant a unique value for standard distance from the mean, a measure that indicates how compact or dispersed the GPS features are. A participant who is primarily in one location (generally home) and stays there will have a much smaller distance from the mean (more compact cluster) than someone who has a much more extensive range over the course of a week (less clustering). Significant work has been done on clustering in spatial datasets including variance across scales of analysis (R. Wilson & Din, 2018), sociological and cultural clustering effects (Harold & Fong, 2018), efficient analysis of very large scale spatial datasets (Ester et al., 1996) based on range of clustering algorithms (J.-S. Chen et al., 2004). In this study, the details of specific location-based activities is not a significant interest making a single clustering feature sufficient.

The formula used to calculate the standard distance from the mean is given as a value over (n) features where x_i and y_i represent a specific GPS reading (i) and $\{\bar{X}, \bar{Y}\}$ is the mean centre of the given features (Mitchell, 2005):

Equation 2:

$$Standard\ Distance = \sqrt{\frac{\sum_{i=1}^{n}(x_i - \bar{X})^2}{n} + \frac{\sum_{i=1}^{n}(y_i - \bar{Y})^2}{n}}$$

3.6.1.3 Difference of Median and Mean Distance

A common test for outliers in traditional statistics is to observe the differences in value between the mean and median of a given set of numbers. In the case of a sequential set of features such as those generated in this study by time-ordered GPS readings, the median is calculated by counting the number of features and then identifying the middle value (Figure 3-6).

Figure 3-6 - Median value (green) for a given set of geographic features. In this example with ten features, the fifth value is the median (upper element). If the line is stretched out, the logic of the process becomes clear (lower element). Author designed graphic.

For each of the 97 participants a mean and median were generated and the differences were compared. This produces the variable *difference between mean and median*. This value indicates another variability in the characteristics of a given track that reflects sensitivity to outliers.

3.6.1.4 Rotation of the Standard Distance Ellipse

In addition clustering, dispersal, and outliers, a given set of features may exhibit directional tendencies. For example, someone walking directly from south to north would generate a set of features that is highly oriented North-South with no deviation East or West visually expressed as a highly elongated ellipse along the North-South axis. In contrast, if someone walked evenly back and forth in all cardinal directions, there would be very little notable directionality and the shape of the ellipse would look like a circle where North-South and East-West orientations are equal. The directional value of a set of features can be calculated by comparing the rotation from zero degrees (North) each line segment or vector is (Figure *3-7*).

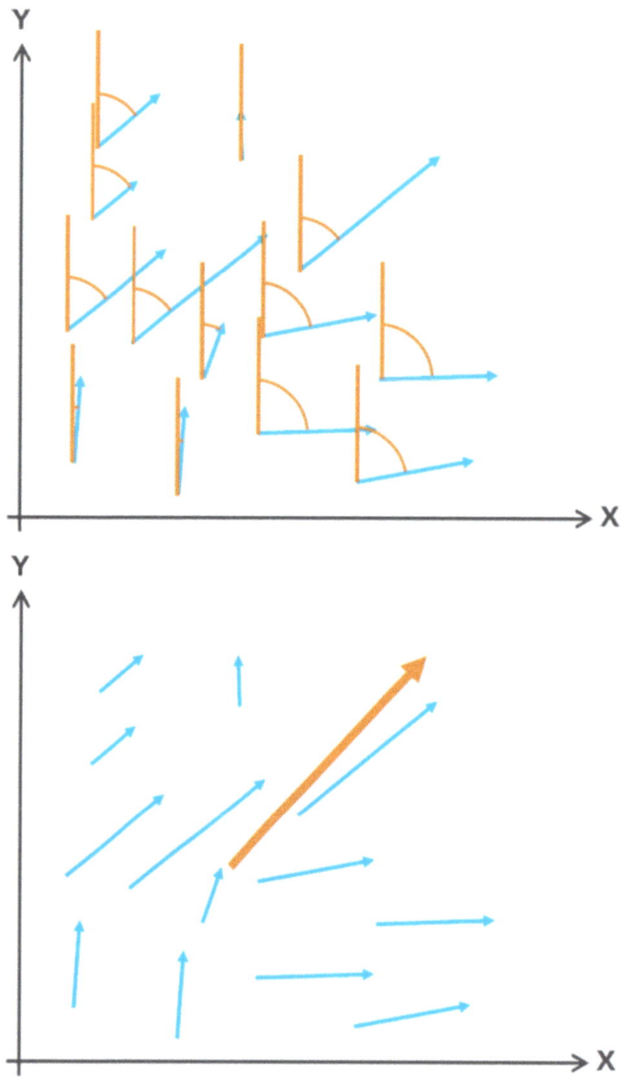

Figure 3-7 - Directional orientation, or linear mean direction, of a given set of features. Figures are modifications from ESRI technical workshop graphics (Bennett and Vale, 2015).

If these are summed across all vectors in the feature set, a single value can be obtained in the form of a degree reading that provides directional orientation (Cordeiro & De Paula Ferrari, 1991; Mardia & Jupp,

1999). The directional ellipse provides insight about the intensity of the directional tendency through the relative elliptical or circular nature of the resulting standard deviation.

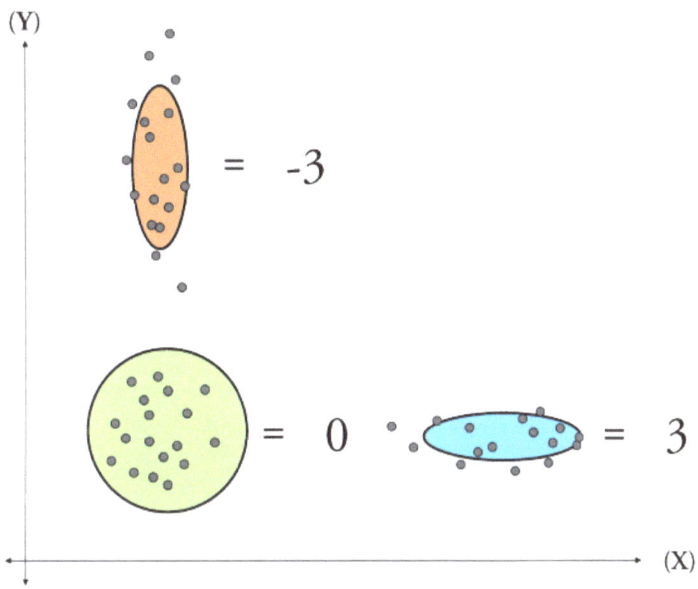

Figure 3-8 - Examples of XY ratio of ellipse measurements.

The calculations used to arrive at the values for the directional ellipse and its cardinal orientation are made through a series of steps that begin with individual calculation of the mean centre of the set of features (Gong, 2012) (Figure *3-8*). The angle of rotation is then calculated based on distances from the mean for each x/y coordinate, which is then used to calculate the ellipse which will appear elongated where there is variance between x and y axes (Equation 3):

Equation 3:

$$SDE_x = \sqrt{\frac{\sum_{i=1}^{n}(x_i - \bar{X})^2}{n}} \qquad SDE_y = \sqrt{\frac{\sum_{i=1}^{n}(y_i - \bar{Y})^2}{n}}$$

Angle of rotation:

$$\tan\Theta = \frac{A+B}{C}$$

$$A = \left(\sum_{i=0}^{n} \tilde{x}^2 - \sum_{i=0}^{n} \tilde{y}^2\right)$$

$$B = \sqrt{\left(\sum_{i=1}^{n} \tilde{x}^2 - \sum_{i=1}^{n} \tilde{y}^2\right)^2 + 4\left(\sum_{i=1}^{n} \tilde{x}_i\tilde{y}_i\right)^2}$$

$$C = 2\sum_{i=1}^{n} \tilde{x}_i\tilde{y}_i$$

Standard deviation of x and y-axes:

$$\sigma_{x=} \sqrt{\frac{\sum_{i=1}^{n}\left(\tilde{x}_i \cos\Theta - y_i \sin\Theta\right)^2}{n}}$$

$$\sigma_{y=} \sqrt{\frac{\sum_{i=1}^{n}\left(\tilde{x}_i \sin\Theta - y_i \cos\Theta\right)^2}{n}}$$

The ellipse features and directional orientation identify the aggregate characteristics of significant numbers of individual GPS features.

3.6.1.5 Ratio of X and Y Axes for Directional Ellipse

In addition to cardinal directionality expressed in degrees clockwise from North, the standard deviation from the mean of the x-axis and the standard deviation from the mean of the y-axis can be compared to determine how flat or circular the resulting ellipse is quite apart from cardinal orientation. This is a derived, experimental variable developed for this paper as a means of identifying meaningful differences in directional tendency from one participant to another. A perfect circle would yield a value of zero. Values above zero indicate elongation along the x axis while values below zero indicate elongation along the y axis (*Figure 3-9*).

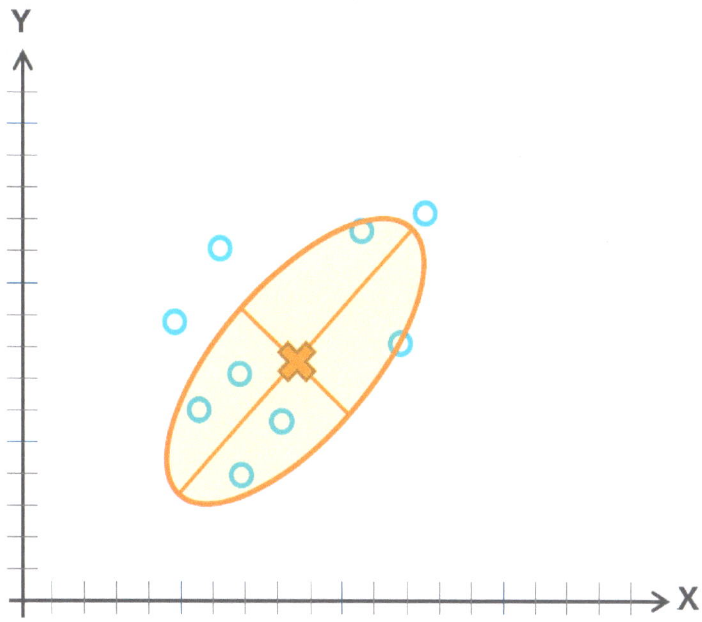

Figure 3-9 - Determination of ellipse through calculation of x and y-axis standard deviations from the mean. Figures are modifications from ESRI technical workshop graphics (Bennett and Vale, 2015).

3.6.1.6 Nearest Neighbour Observed Mean Distance

I begin this section with an important clarification. The use of "neighbour" here does not indicate a human neighbour and has no direct social implications. The term neighbour refers to the proximity of one GPS feature to another. By measuring the proximity of features to each other, various neighbour scores can be produced from the GPS data. This becomes another means to understand the spatial characteristics of a set of GPS features such as the relative compactness of a given set of features or segments of features. Rather than using time-organized linear features, the nearest neighbour calculation treats each feature as a point scattered in a landscape which can be located by reference to other nodes in the network (Yao et al., 2017). If there are 100 features, the expected average is calculated by evenly distributing the 100 elements within the bounds of a box enclosing those 100 features. It then calculates where the actual features are spatially before comparing that with the expected distribution. This produces a value which, if it is greater than 1, indicates a scattered set and if less than 1, a clustered set (Figure 3-10).

The focus on finding patterns in the GPS data means that individual trips with noted beginning and end points, home-work trips, home-recreation or similar detailed movement are not examined. For a neighbourhood scale spatial-social correlation study, large numbers of aggregated GPS data points are more critical than specific detail for any one trip or segment made by an individual participant (R. M.

Carpiano, 2006; Savitz & Raudenbush, 2009; Yao et al., 2017) much as the specific position or movement of a gas molecule is not relevant to a measure of pressure for a gas in an enclosed space.

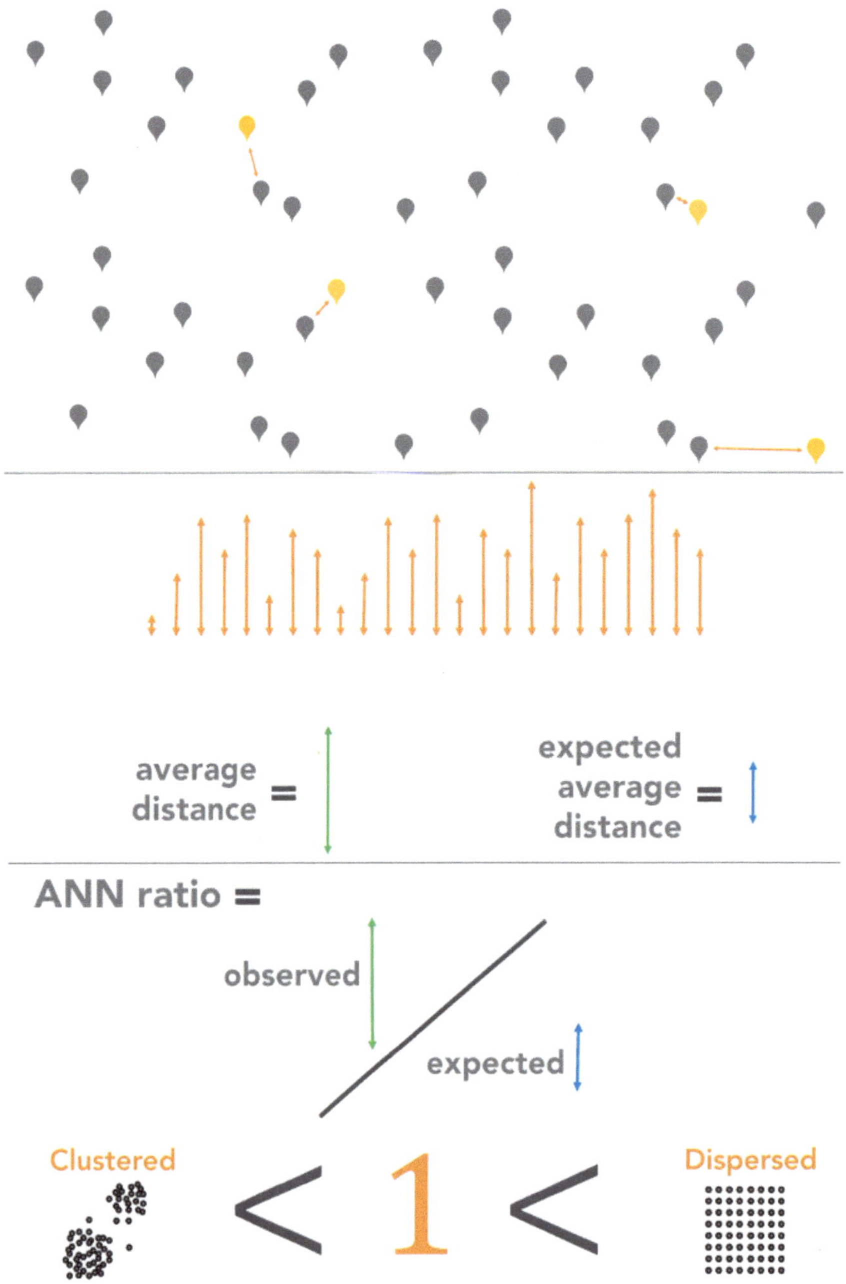

Figure 3-10 - Nearest neighbour conceptual description. Figures are modifications from ESRI technical workshop graphics (Bennett and Vale, 2015).

Within networks, whether spatial, social, natural or technological, the degree of clustering is an important feature that warrants measurement (Blondel et al., 2008; J.-S. Chen et al., 2004). The means by which these values are determined arises from an application of the following nearest neighbour calculations:

Expected Mean Distance where A is total study area, n is the total number of features.

$$\overline{D}_E = \frac{0.5}{\sqrt{\frac{n}{A}}}$$

Observed Mean Distance where d_i is the distance between feature *i* and the next closest feature.

$$\overline{D}_O = \frac{\sum_{i=1}^{n} d_i}{n}$$

Once calculations for the variance between expected distance and observed distance are made, the calculation for the average nearest neighbour is possible.

Average Nearest Neighbour where D_o is observed distance and D_E is expected distance.

$$ANN = \frac{\overline{D}_o}{\overline{D}_e}$$

3.6.1.7 Nearest Neighbour Z-score

If the Z-score for the nearest neighbour values are calculated by solving for the Standard Error, it is possible to determine the extent to which the given features are randomly distributed (Ebdon, 1976, 1991; R. Wilson & Din, 2018). We can calculate the Z-score for Average Nearest Neighbour where SE is the Standard Deviation by using:

$$z_{ANN} = \frac{\overline{D}_O - \overline{D}_e}{SE} \qquad SE = \frac{0.26136}{\sqrt{\frac{n^2}{A}}}$$

3.6.1.8 Nearest Neighbour Area

This variable uses the A variable above which represents the total study area. With networks, we can speak of network diameter (Brandes, 2001) to indicate how spread out a network is . This characteristic is sometimes understood in geo-statistical analysis as spatial segregation (Wong, 1999). In spatial statistics, a value called *nearest neighbour area* can be determined by calculating the area of a rectangle which covers all of the features in a given spatial data set – in this case GPS observations (Ebdon, 1991). The value reflects a slightly different behaviour marker than total length of a GPS path because total length of a path does not indicate whether that travel covers a large geographic area – e.g. Driving around a racetrack for seven days would yield a very long path length and a very small study area for a given participant (Figure *3-12*).

Figure

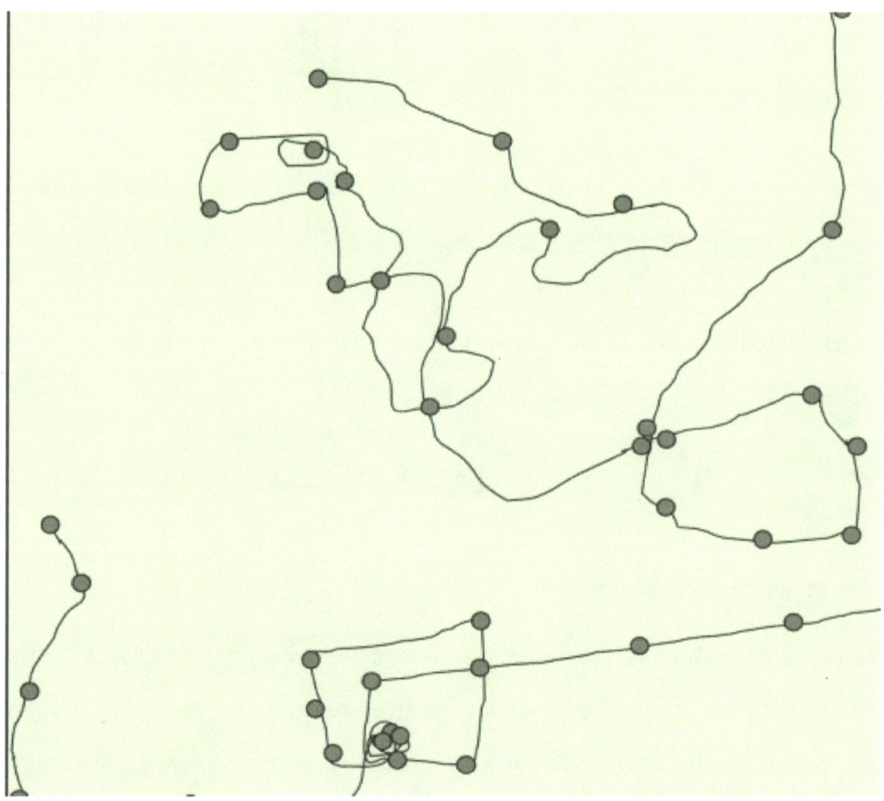

Figure 3-12 - Rectangle enclosing all points used to calculate total area of travel territory.

3.6.1.9 Standard Speed

The calculation of a standard speed for a given set of data points that are time-oriented is based on the mean of the total of all line segments. Velocity of each line segment is determined by calculating the distance between each point and then dividing that by time so that $\boldsymbol{Velocity} = \frac{Distance}{Time}$ for each segment. The value is totaled and then divided by the number of segments to determine the mean velocity for the time-series data points. This velocity is referred to as the *standard speed* measure reflects a synthesis of modal travel types (none, walk, vehicle) as an average speed across the GPS track.

3.7 Results of Spatial Analysis

A summary of results for each of the ten spatial variables clustered by Census Tract divisions is provided below. This strategy is in keeping with the study design which is examining how spatial use patterns may be correlated with demographic and social data examined earlier. Rosedale(Mid) has the shortest total distance when all the tracks are summed (Figure *3-13*).

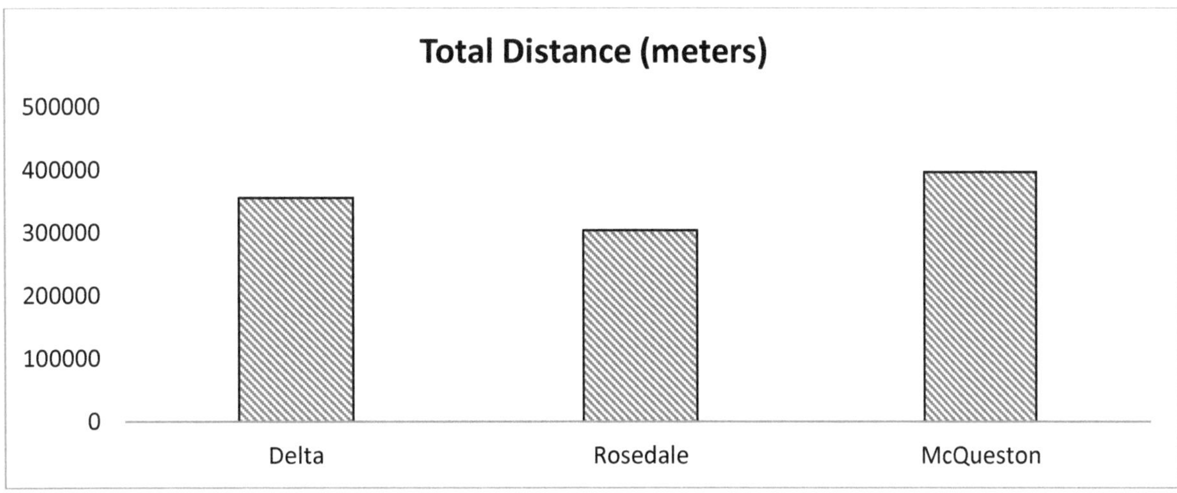

Figure 3-13 - Comparison of total distance of GPS track measured by calculating distance between each reading (unweighted).

The McQueston(Low) has the longest total distance by mean average. Delta(High) has a significantly greater standard distance radius, Rosedale(Mid) has a shorter standard distance than McQueston(Low) so income doesn't appear to be a determining factor in how widely dispersed the GPS data points are at a Census Tract level (Figure *3-14*).

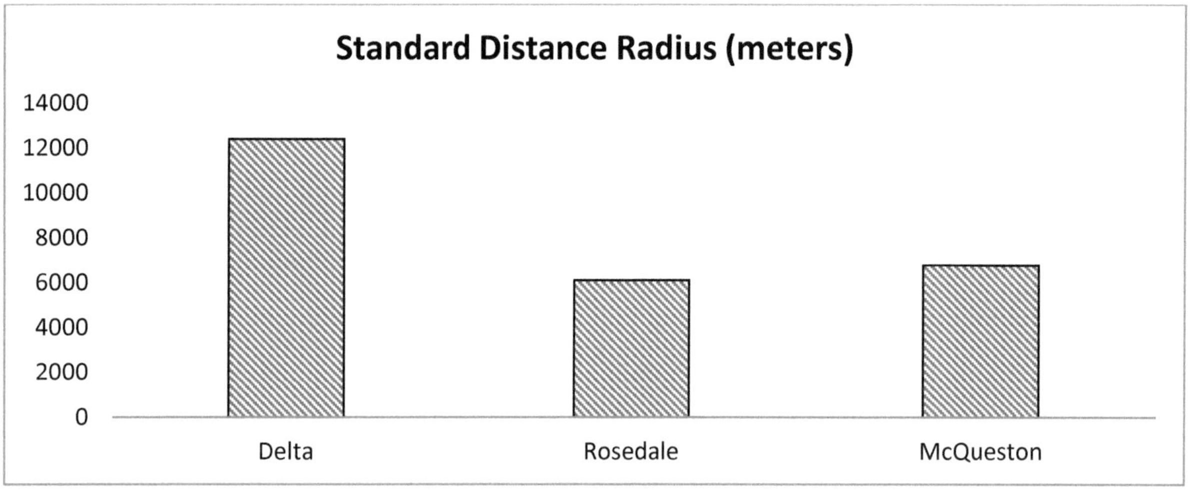

Figure 3-14 - Size of circle drawn around the mean centre determined by standard deviation from the mean.

The geographic distance between mean location and median feature is significantly lower in Rosedale(Mid). The McQueston(Low) mean/median difference is five times greater than Rosedale(Mid). The Delta(High) is approximately halfway between these two values (

Figure 3-15).

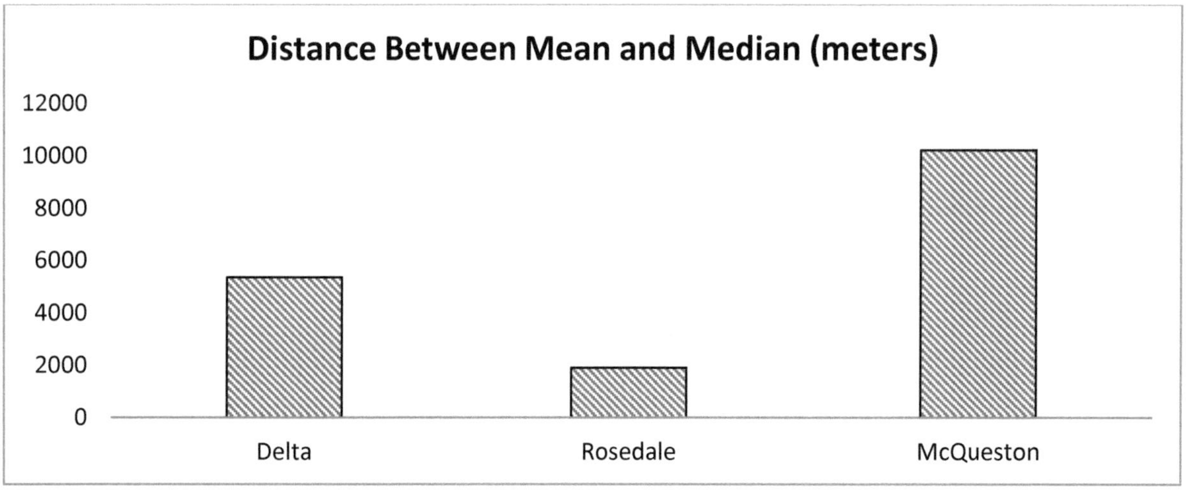

Figure 3-15 - Comparison of three Census Tracts using the difference between mean center and median center.

Greater mean/median distance likely signifies the presence of a greater number of outliers. When we compare total distance, this same neighbourhood has only a slightly longer mean path length. This further suggests that the presence of outliers in the lower income neighbourhood is limited to a very small number of cases.

There is a steady but modest increase in the Easterly rotation of the directional ellipse from the Delta(High) to the lower income neighbourhood. Rotation is measured clockwise from North so a rotation of 90 degrees would be oriented in an Easterly direction. Delta(High)'s average rotation is 83 degrees, Rosedale(Mid)'s is 86 and McQueston(Low)'s is 95 degrees (Figure 3-16).

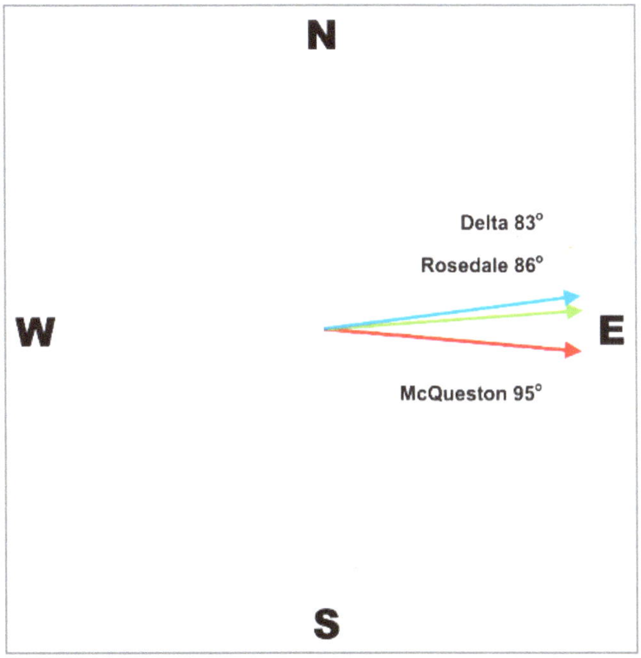

Figure 3-16 - Deviational Ellipse Rotation. Comparison of the degree of rotation of the standard deviational ellipse across three Census Tracts(degrees clockwise from North). Rotation is 3 degrees more easterly in the middle income neighbourhood than in the Delta(High). The lower income neighbourhood is 12 degrees more easterly than the Delta(High).

The ellipse in the Rosedale(Mid) is the most stretched in an east-west direction. The Delta(High) is tends more toward a round shape while the lower income neighbourhood occupies a middle position between them. The significance of the shape is likely related to the urban form of the city and suggests that urban form variations between the study areas is detectable but not significant (Figure 3-17).

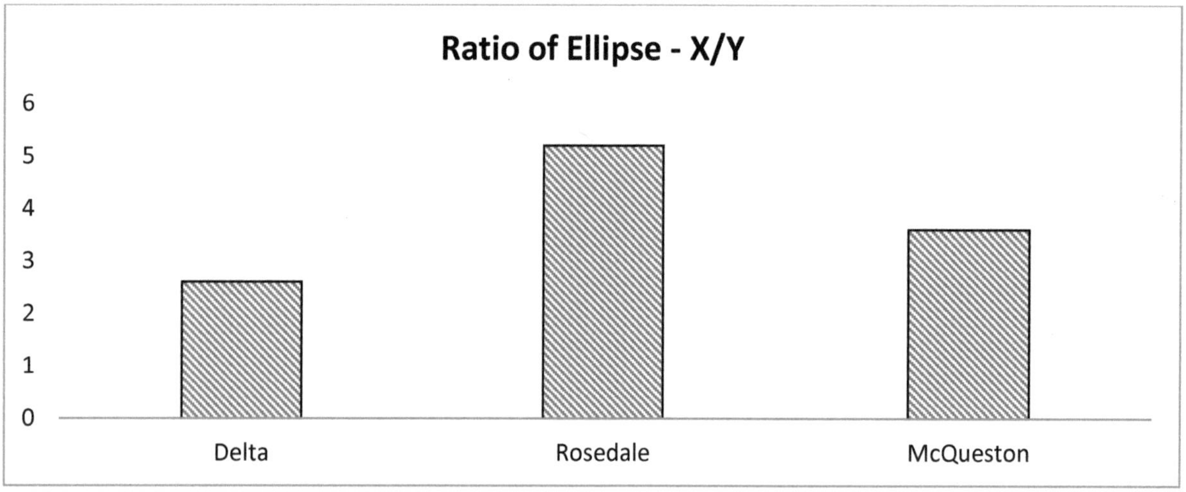

Figure 3-17: Ratio of the X and Y axes showing relative degree of roundness

Primary transportation arteries that are supra-neighbourhood and which carry the highest traffic flows to and from these neighbourhoods are oriented in an East-West tendency. The middle income neighbourhood is particularly influenced by an East-West artery on its North boundary owing to its triangular shape where the other two sides of the triangle are a natural feature that prevents vehicular navigation (the Niagara Escarpment) and a freeway structure with limited access (Red Hill Parkway) (see Figure 2-4: The white circle in the top image is a 2km radius around the center point of the three core Census Tracts. The lower map denotes the primary boundaries of the study area: Ottawa Street to the West, Barton Street to the North, the Red Hill Expressway to the East and the Escarpment to the South.

earlier). The Delta(High) is the most symmetrical on all sides with North-South and East-West travel lanes of significance on all four sides. The lower income neighbourhood has primary access on an East-West artery and a North-South artery on one side (West). Given these features, we would have expected that if vehicular transport is primary, the Rosedale(Mid) would be the most stretched East-West. This is in fact the case with the Delta(High) stretched the least with the lower income neighbourhood stretched to a value between them. This is also the case.

The data suggests that the lower income neighbourhood is the most clustered and the Rosedale(Mid) is the most spread out in terms of travel patterns (Figure 3-18).

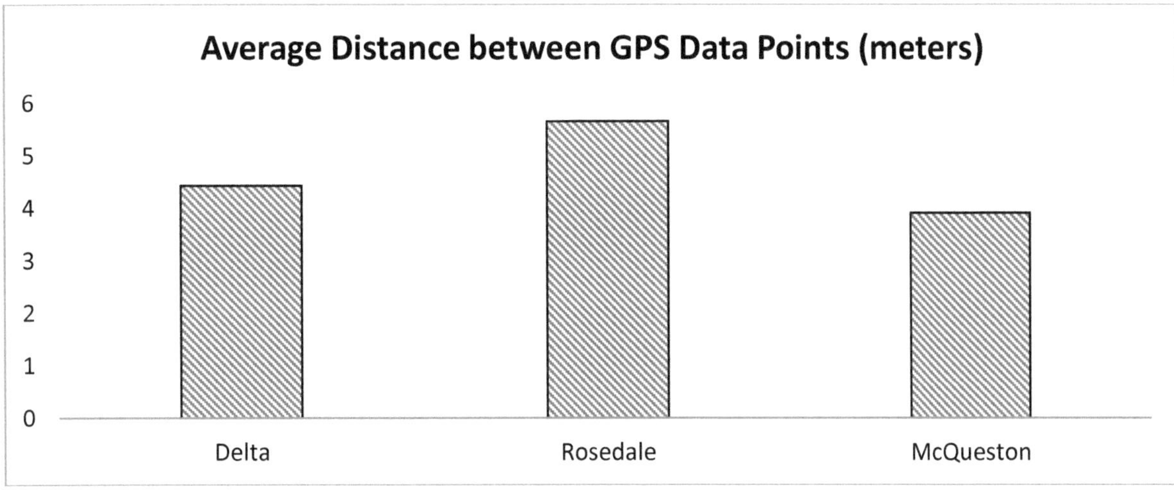

Figure 3-18 - Average mean distance between GPS data reading points.

When the tracks are measured using point-to-point distance and compared with a random distribution of points, we can determine how likely it is that the human-produced tracks are random. A value of less than 1 means clustered and not random while a value higher than 1 means dispersed and more random (Figure *3-19*).

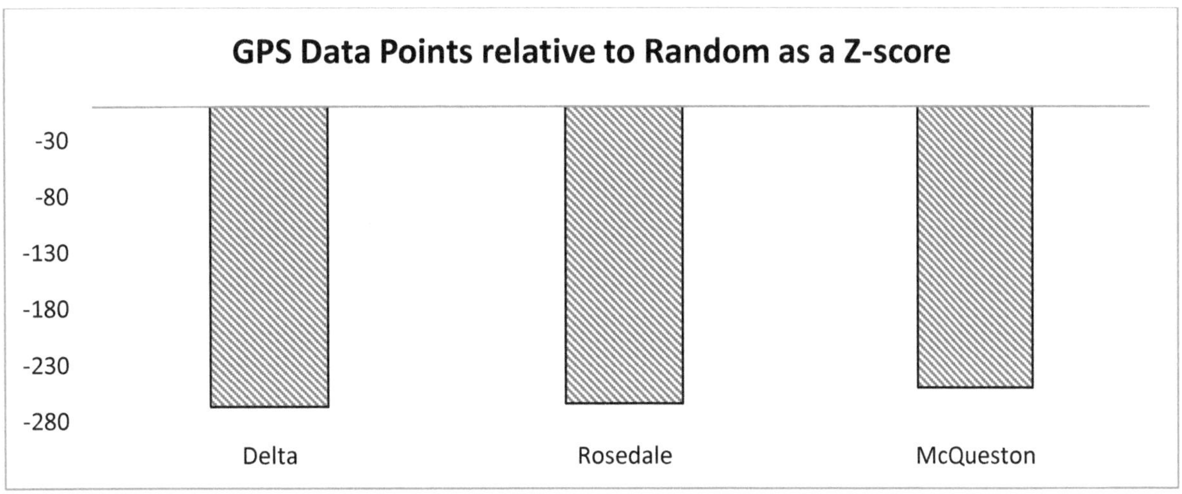

Figure 3-19 - Comparison of Z-scores across three Census Tracts.

Analysis of the participant data tracks reveals a very high clustering which should also be reflected in the Z-score. The tracks are not random as has been noted in other spatial research (Christou, 2013). The differences between the high and low value for the three Census Tracts is just slightly less than 1% (Figure 3-20). Random human movement patterns are not the concern of this study but a baseline of randomness provides a contrast for the very non-random nature of human movement even when aggregated across all 97 participant's tracks. As comparison and analysis is carried out on spatial movements of this kind, we can be confident that GPS track patterns are not at all a random element.

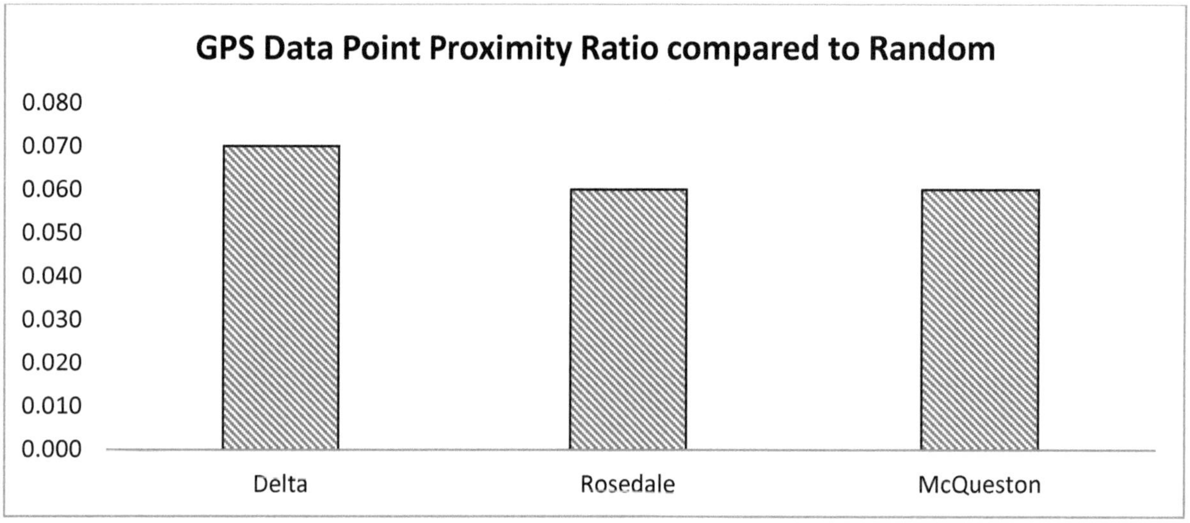

Figure 3-20: Comparing degree of randomness in point to point ratios across three Census Tracts.

This graph suggests that Delta(High) is the most clustered and McQueston(Low) is the least clustered. The values, however, expressed as a percentage show that the relative difference given the large overall value is very small ~ 6 difference between the top and bottom value.

Another behavioural characteristic is the size of the geographical area encompassed by a rectangle that covers all of the GPS points for a given track. The notably larger average area of Delta(High) suggests that despite the possible presence of outliers from earlier graphs for the lower income neighbourhood, the territory covered by higher income participants is about 15% larger than the lower income neighbourhood and 26% larger than the Rosedale(Mid) (Figure *3-21*).

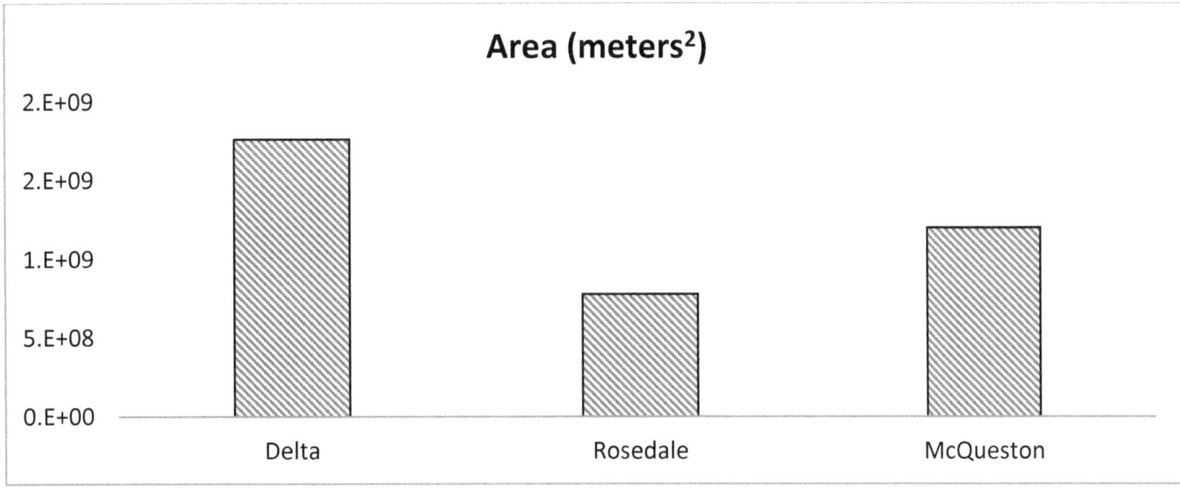

Figure 3-21 - Area of rectangle drawn around all GPS data points - mean average of all tracks per Census Tract.

Average higher standard speed distributions would suggest a higher rate of vehicular travel and/or greater use of higher speed connections – e.g. In-city highways or primary arteries rather than slower neighbourhood grid street travel. Modal details at a participant level are not the focus of this study but average speed is a spatial movement characteristic that is affected by a behavioural decision to stay, walk, bike, drive, and so on. Modal characteristics are summarized below. The variance between the values averaged over each Census Tract are marginal with 0.89% between maximum and minimum values (Figure *3-22*).

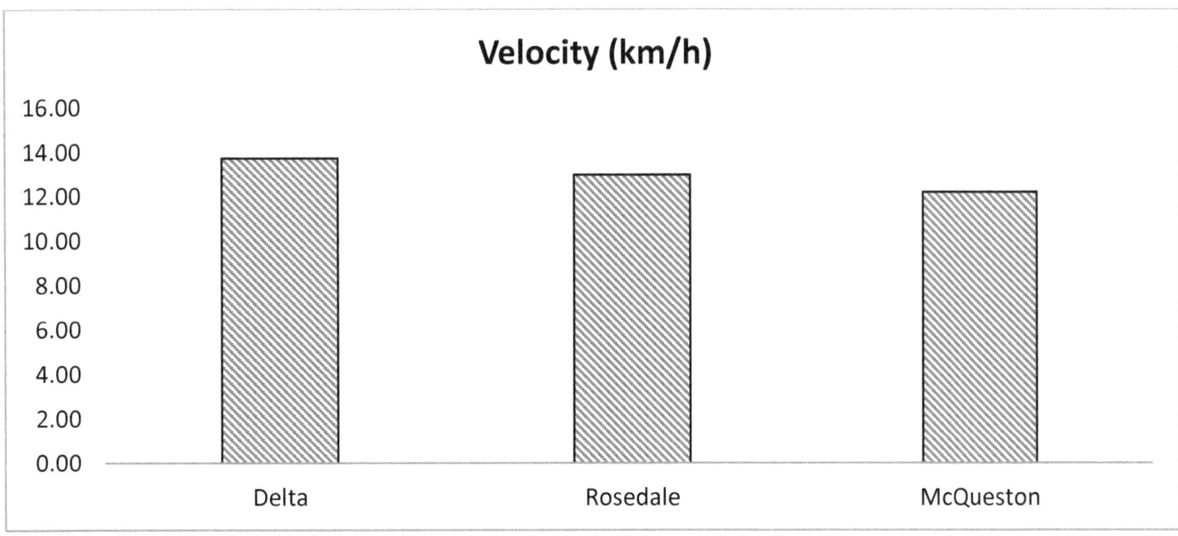

Figure 3-22 - Standard distribution of the mean value of "speed" variable across all GPS data points by Census Tract.

3.8 Modal Analysis

Travel modes (walk, bike, bus, car, etc.) are not a primary focus of this paper but modal patterns represent an important aspect of the description of human spatial patterns in cities. Large-scale or aggregated averages (e.g. values in Figure *3-22)* tend to mask modal details of specific individuals. To that end, an analysis of modes has been undertaken. Modal analysis provides insight into the types or modes of travel that participants engaged in while logging location data. Given the significant complexity of distinguishing very similar modes such as bus and car on city streets with shared right of way including common traffic signals (Huss et al., 2014).

A more coarse-grained segmentation and classification strategy for reviewing modal characteristics was used for this study. Analysis was based on *speed* column data, a value arrived at by calculating the Euclidean distance between GPS data points logged at 15 second intervals. A simple modal detection calculation is used based on machine learning that has established the mean speeds of various modes at given time intervals (Gong, 2012; Huss et al., 2014; Omrani, 2015; Omrani et al., 2013; L. Zhang et al., 2011). For each of the 97 tracks, average (mean) speed is calculated over 195 second intervals (15 seconds x 13 readings = 195 seconds) to produce a value. That value is then assigned to a modal range where 0-2km/h is equal to *None*, 2-8km/h is equal to *Walk*, 8-48km/h is equal to *City* (denoting vehicle travel on city streets) and 48-120km/h is equal to *Highway* (denoting travel speeds well above normal city street average values). This value is then assigned to a column representing mean speed data and sorted into bins via a histogram function giving respective modes: none, walking, city vehicle travel and highway vehicle travel (L. Zhang et al., 2011).

After processing the 97 Social Imaging GPS data tracks (two million one hundred thousand line segments), the following results were obtained (Figure 3-23):

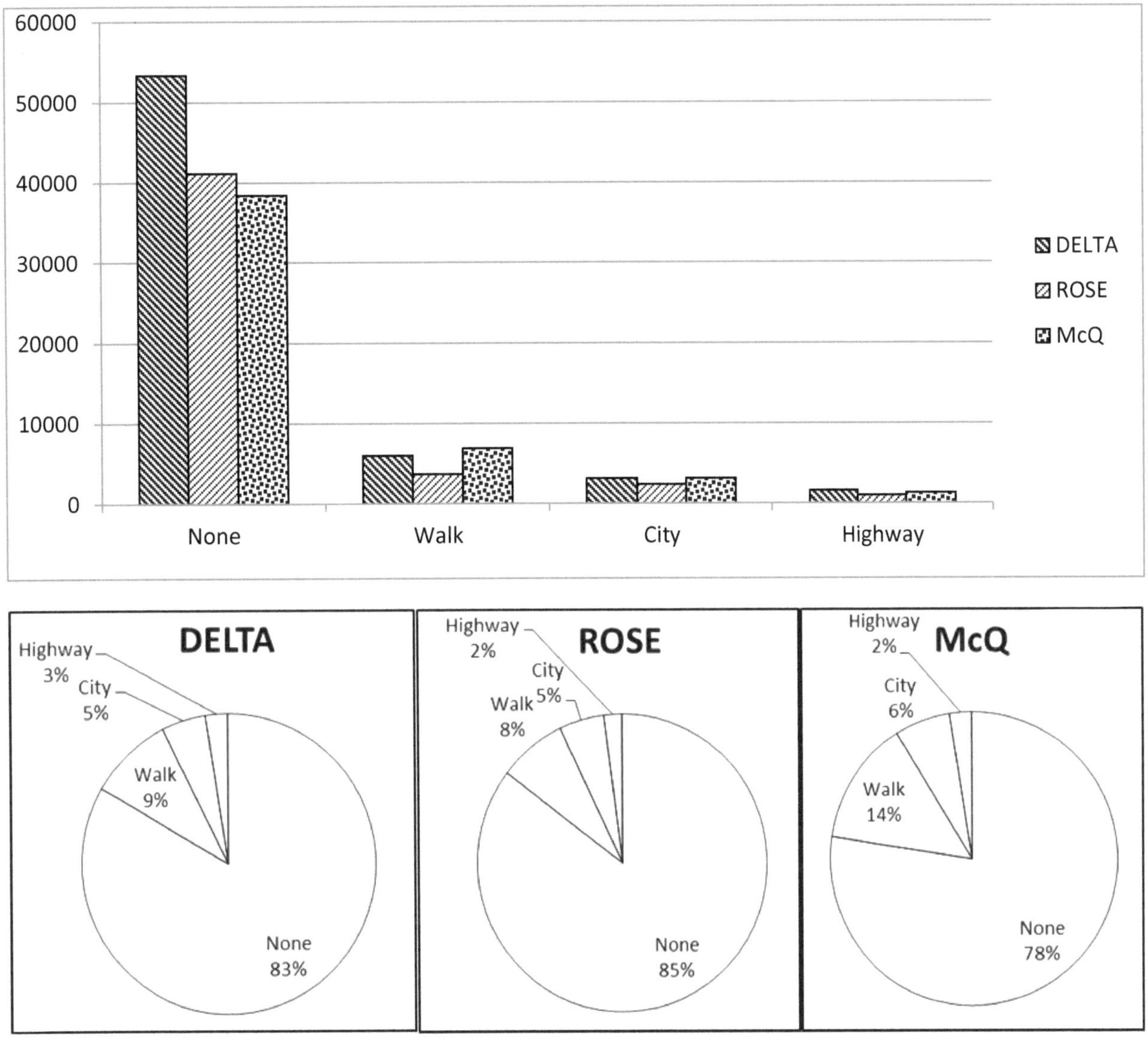

Figure 3-23 - Results of modal analysis of social imaging data by neighbourhood. "None" in both bar graphs and pie charts means no movement.

Overall the patterns show similarity with the *None* category making up a large majority of travel segments. Highway use was also relatively common across the three Census tracts. Delta(High) had 1% more *Highway* travel than Rosedale(Mid) or McQueston(Low). *City* segments were also similar with McQueston(Low) City travel at 6%, Delta(High) at 5% and Rosedale(Mid) at 5%. The two most notable variances were *Walking* where that mode comprised 14% of segments for McQueston(Low), 9% for Delta(High) and 8% for Rosedale(Mid). The accumulation of differences between *Walking*, *City* and *Highway* across all three geographic divisions led to an inactivity level for Rosedale(Mid) (None) of 85%, Delta(High) of 83% and McQueston(Low) of 78%. Based on these results, we can conclude that

McQueston(Low), showed higher *Walk* and *City* levels and the lowest levels of *None* at 78% compared with 83% for Delta(High) and 85% for Rosedale(Mid).

3.9 Discussion of Demographic and Spatial Comparison

The following chapter will formally examine the relationship (if any) between social capital and spatial metrics but in the interest of examining as far as possible any relationships in the data between spatial patterns and the participants in the survey, a review of the correlations between six spatial features and the 15 demographic and context variables is undertaken in brief below. After a correlation matrix of all ten spatial and 15 demographic and context variables was generated, six are singled out for review below. The intent of this study remains focused on the central and secondary hypothesis. What follows are additional observations about variances across the study areas. These in turn are considered for potential relations and insights that could be linked to the spatial and social capital analysis undertaken in Chapter 4.

The nature of the spatial data generated by GPS data points lends itself to a wide range of analyses. The largely descriptive nature of the foregoing exploration has yielded preliminary signals and patterns about which we may make the following observations though the possibilities for examining this large spatial data set have not been exhausted. The data suggests the following.

1. *Delta(High) had a greater standard distance radius when compared with the other two study areas. This appears to reflect a wider range of travel outside the Census Tract when averaged over all participants in that neighbourhood. Additional correlations reflect other possible relationships such as a positive correlation between income and distance for Delta. This could provide support for the central hypothesis should that hypothesis be supported in the following chapter (*Figure 3-24*). It is not clear why there is negative correlation between years in dwelling and neighbourhood with distance for McQueston but could suggest longer tenure leads to smaller travel distance.*

Q1:		age	gender	marital.status	religion	income	education	dwell.type	years.dwell	years.neighbourhood	number.neighbours	belonging	primary.activity	vote.freq	volunteer	donate
DELTA	standis.radius	-0.06	0.17	-0.12	0.05	0.35	0.13	-0.12	0.07	-0.01	0.01	0.01	-0.22	0.1	0.16	0.06
ROSEDALE		-0.31	0.19	0.01	-0.2	0.2	0.28	-0.18	-0.03	-0.02	0.23	-0.25	-0.35	0.13	0.25	0.11
MCQUESTON		0.27	0.46	0.17	-0.08	0.23	0	0.26	-0.43	-0.47	-0.15	0.02	0.04	-0.02	-0.2	-0.16

Figure 3-24 - Correlation table for Standard Distance Radius and Demographic variables.

Total distance was greatest for McQueston(Low). When compared to standard distance radius, it would suggest the possibility of more long distance paths or more participants travelling longer distances each but not necessarily as many long distance tracks as Delta(High) – i.e. More participants travelling mid-range distances. Additional correlations reflect other possible relationships (Figure 3-25).

Q2:		age	gender	marital.status	religion	income	education	dwell.type	years.dwell	years.neighbourhood	number.neighbours	belonging	primary.activity	vote.freq	volunteer	donate
DELTA	total distance	-0.17	0.18	-0.12	-0.05	0.3	0.17	-0.1	0.08	-0.02	0	-0.03	-0.29	0.03	0.13	0.09
ROSEDALE		-0.18	0.03	-0.04	-0.15	0.12	0.15	-0.19	-0.17	-0.14	0.13	-0.1	-0.42	-0.11	0.34	0.18
MCQUESTON		0.26	0.49	0.26	-0.14	0.26	0.19	0.14	-0.2	-0.24	-0.17	0.02	0.03	-0.18	-0.12	-0.28

Figure 3-25 - Correlation table for Total Distance and Demographic variables

2. *This is mitigated, however, by a greater distance between Median and Mean of the McQueston(Low) Census Tract. Normally, this would suggest more outliers for McQueston(Low) but the larger standard distance radius for Delta(High) seems to confound this. It may be that Delta(High) has an average long distance dispersal greater than McQueston(Low) or that Delta(High) has a low number of very significant outliers. Additional correlations reflect other possible relationships (Figure 3-26).*

Q3:		age	gender	marital.status	religion	income	education	dwell.type	years.dwell	years.neighbourhood	number.neighbours	belonging	primary.activity	vote.freq	volunteer	donate
DELTA	dif.memed	-0.14	0.18	-0.1	0.05	0.37	0.04	-0.12	0.03	-0.06	0.13	0	-0.29	0.1	0.18	-0.02
ROSEDALE		-0.35	0.13	0.16	-0.05	0.15	0.04	-0.1	0.01	0	0.15	-0.29	-0.24	0.16	0.09	0.01
MCQUESTON		0.13	0.14	0.27	0.04	-0.09	-0.02	0.57	-0.54	-0.56	0.11	-0.09	0.08	0.22	-0.26	0.03

Figure 3-26 - Correlation table for Difference of Mean / Median Centre and Demographic variables

3. *McQueston(Low) has a slightly stronger East-West orientation when the directional ellipse is considered. This suggests that there are longer trips in the East-West direction or more trips more closely aligned to the East-West axis. The urban form appears to be a strong contributor to this as the lower City in Hamilton is strongly constrained East-West by the industrial and harbour areas to the North and the Niagara Escarpment to the South. Additional correlations reflect other possible relationships (Figure 3-27).*

Q4:		age	gender	marital.status	religion	income	education	dwell.type	years.dwell	years.neighbourhood	number.neighbours	belonging	primary.activity	vote.freq	volunteer	donate
DELTA	ellips.rotat	0.12	0.04	-0.4	0.18	-0.07	-0.38	0.2	0.06	-0.01	0.01	-0.09	-0.01	-0.04	-0.05	-0.04
ROSEDALE		0.29	-0.14	-0.2	0.16	-0.01	-0.09	-0.03	0.24	0.22	0.01	0.01	0.15	0.12	-0.25	-0.05
MCQUESTON		0.32	-0.08	0.12	-0.22	-0.09	0.02	-0.02	0.23	0.27	0.11	-0.2	0.17	0.03	-0.16	-0.01

Figure 3-27 - Correlation table for Ellipsis Rotation and Demographic variables

4. *Directional orientation related to urban form is supported by the X/Y ratio measures. Delta(High) has the most rounded form and Rosedale(Mid) has the most elliptical form. Delta(High) has grid and artery road access on all sides of the Census Tract, McQueston(Low) has access on three sides, and Rosedale(Mid) has access on one side (being constrained by a Red Hill freeway and creek on one side and the Niagara Escarpment on the other). Additional correlations reflect other possible relationships (Figure 3-28).*

Q5:		age	gender	marital.status	religion	income	education	dwell.type	years.dwell	years.neighbourhood	number.neighbours	belonging	primary.activity	vote.freq	volunteer	donate
DELTA	ellips.xyrat	0.19	0.21	-0.26	0.03	0.18	-0.21	-0.12	0.1	0.03	-0.21	-0.05	-0.01	0.09	-0.05	-0.21
ROSEDALE		0.18	0.07	-0.26	0.08	-0.44	-0.09	0	-0.35	-0.34	0.01	0.21	0.11	-0.38	0.01	0.11
MCQUESTON		0.14	0.45	0.23	-0.19	0.3	0.18	-0.12	0.12	0.15	-0.08	0.01	-0.06	-0.2	-0.07	0.18

Figure 3-28 - Correlation table for Ellipsis X and Y Axes Ratio and Demographic variables

5. *Delta(High) also shows a slightly higher nearest neighbour clustering on both the Nearest Neighbour Ratio and Nearest Neighbour Z-score. This indicates a more clustered neighbourhood than the other two. McQueston(Low) had slightly less clustered profile suggesting a slightly more dispersed spatial network. Additional correlations reflect other possible relationships (
Figure 3-29).*

		age	gender	marital.status	religion	income	education	dwell.type	years.dwell	years.neighbourhood	number.neighbours	belonging	primary.activity	vote.freq	volunteer	donate
Q6A: Delta	nn ratio	-0.05	-0.13	0	-0.02	-0.07	-0.07	-0.06	0.1	0.06	0.06	0.05	-0.14	-0.32	-0.14	-0.31
Rosedale		-0.31	-0.06	-0.08	-0.09	0.17	0.01	-0.11	0.04	0.02	-0.27	-0.2	-0.05	-0.05	-0.19	-0.22
McQueston		-0.17	-0.16	0.1	0.18	-0.19	-0.12	0.27	-0.09	-0.09	0.28	-0.05	0.14	0.33	-0.3	-0.06

		age	gender	marital.status	religion	income	education	dwell.type	years.dwell	years.neighbourhood	number.neighbours	belonging	primary.activity	vote.freq	volunteer	donate
Q6B: Delta	nn zscore	0.02	0	-0.03	0.23	0.04	-0.13	-0.12	0.25	0.13	0.35	-0.12	0.11	0.05	0.08	0.15
Rosedale		-0.36	-0.09	-0.2	-0.04	0.22	-0.08	-0.23	0.13	0.13	-0.21	-0.36	-0.13	0.23	0.08	0.05
McQueston		-0.14	-0.04	0.24	0.09	-0.2	-0.13	0.45	-0.12	-0.12	0.23	-0.11	0.18	0.36	-0.35	-0.15

Figure 3-29 - Correlation tables for Nearest Neighbor Ratio and Nearest Neighbor Z-Score and Demographic variables

6. *Delta(High) also had a significantly larger overall range (area of rectangle around all points) and slightly higher average speed distribution. These two factors appear to be related given that wider travel ranges entail use of higher speed roadways and highways. We may surmise from this that vehicular travel is highest in Delta(High) but only marginally so. Additional correlations reflect other possible relationships (Figure 3-30).*

		age	gender	marital.status	religion	income	education	dwell.type	years.dwell	years.neighbourhood	number.neighbours	belonging	primary.activity	vote.freq	volunteer	donate
Q7: Delta	allpointsarea rectangle	0.01	0.23	-0.16	0.07	0.27	0.13	-0.1	0.12	0.07	-0.14	-0.03	-0.18	0.07	0.18	0.03
Rosedale		0.08	0.18	-0.21	-0.08	0.06	0.4	-0.14	0.08	0.12	0.46	-0.16	-0.25	0.08	0.38	0.25
McQueston		0.21	0.41	0.25	-0.14	0.01	0.23	0.32	-0.21	-0.24	-0.07	-0.07	0.24	-0.1	-0.07	-0.27

Figure 3-30 - Correlation table for All Points Area Rectangle and Demographic variables

Further exploration can be made of the data include elements or comparisons not directly considered here. These dimensions and angles include the following:

1. *Coincident time-space analysis was not done. The purpose of this part of the study was not to look for relationships among participants but to find average metrics in terms of spatial use patterns. An extension of the study, however, cold explore relational probabilities given the time/location data that is available (Crandall et al., 2010).*

2. *Integrating and collecting GPS data points to simplify the line characteristics could yield a different set of measures which could contain meaningful relations not seen in the total GPS track data analysis (remove duplicates within certain clusters, bracket by time of day). However, care must be taken to avoid erasing signal as tracks are cleaned. I judged it more important to retain consistency in processing across all tracks rather than using an external measure of GPS point cleaning that may have unwittingly eradicated weak or subtle signals.*

3. *Constraining paths to streets and subjecting data to analysis with tools such as the Route Choice Analysis (Dorninik Papinski & Scott, 2011) may reveal finer grained travel details such as the difference between vehicular and transit routes, walking or biking portions, etc. I developed the modal analysis above as a more clear and direct way to do a neighbourhood, general pattern analysis of what the Route Choice Analysis tool could do in more detail. Such detail, however, would take this study well beyond the range suggested by the central and secondary hypotheses.*

4. *Diary data could be further analyzed, summarized, and compared to the spatial characteristics noted as well as the demographic, context, social network and trust components of the SCGSS survey instrument. Again, this is beyond the current study. The diary data, however, was useful in determining typical travel patterns to establish a normal baseline, time use at a granularity of six-hour blocks, and in supporting the conclusions of the modal analysis.*

Chapter 4: An Exploratory Approach for Measuring Social Capital by using Spatial Data

4.1 Introduction

The relationships between social and spatial data are analyzed in this chapter using both supervised (linear regression with input predicting output) and unsupervised methods (random forest and principal component analysis looking at patterns and variable importance ranking). The three different approaches maximize the potential to see patterns in the data that a single method cold easily miss. In addition, the methods are applied to a very well-research public data set to demonstrate the how the methods would detect patterns if those exist in the social-spatial data collected for this project.

Research repeatedly demonstrates that weak, insular and disconnected social networks characterizes citizens in contemporary cities with negative health and well-being consequences (Holt-Lunstad et al., 2010; McPherson et al., 2006; Pantell et al., 2013) These realties are experienced locally, by individuals in neighbourhoods, rather than abstractly at national levels (Mehaffy & Alexander, 2016). While this does not eliminate legitimate concerns about commensurability, local information may be more defensible than national averaging (Wendy N. Espeland & Stevens, 1998; Wendy Nelson Espeland & Sauder, 2007). Personal data are more and more available and important for a wide range of research problems (Deng et al., 2018; Eubank et al., 2004).

Advances have been made in relating social and spatial dynamics including the effect of local scale behaviour on larger patterns of urban form and social development (Hillier & Hanson, 1989) including assessments of the significant scholarly growth generated at the intersection of the social and spatial (Mario L. Small & Adler, 2019) and the use of very large sets of socio-spatial data generated by online social platforms (Bailey et al., 2019).

Current government research instruments such as the General Social Survey lack the resolution for local scale analyses. Exploring potential correlations between spatial use and social capital levels at a neighbourhood or Census Tract level may contribute to our understanding of the dynamic social fabric of our cities at higher resolution, functioning as a social MRI of the city that reflects the 'soft tissue' alongside much more well-known X-ray type physical infrastructure data used in GIS software by engineers and planners. Despite advances in digital communications, it appears that human social capacity remains constrained by the limits of time and attention so that we have few very close friends and a wider circle of decreasing intensity somewhere around 150 relationships (Dunbar, 1993, 2011). Distance from work, social, civic and religious settings is known to play a role in social capital formation and conservation (Foster et al., 2019) while greater social capital may play a role in lower levels of poverty (Harrison et al., 2019).

Understanding the social dynamics of neighbourhoods provides a scale within which these smaller human interactions are located. In turn, this insight could positively impact urban planning policies related to the social infrastructure of communities including the potential means to remedy social pathologies such as social isolation (Holt-Lunstad et al., 2010; McPherson et al., 2006), suicide (C.-Y. Lin et al., 2019), eroding civic institutions (Rupasingha et al., 2002) and lower rates of economic growth (Westlund & Adam, 2010). These relationships are not simple and it may also be unclear what kinds of benefits social capital confers and how those might be understood at neighbourhood levels (Mohan et al., 2005).

Formal causal relationships that are generalizable are unlikely to be discovered at small scales given the phenomena being investigated. Social dynamics cannot be isolated in mechanistic lab settings and applied everywhere. Our understanding of the human agency of citizens' means there is a wide margin of stochastic behaviour. These social research challenges are widely known, including the difficulty of identifying fine-grained patterns among small scale groups or individuals when averaging occurs across a larger population (Keuschnigg et al., 2018; Page, 2017; Watts, 2017). This should not, however, dissuade use from seeking greater clarity concerning social capital measurement, including possible proxy measures, at neighbourhood scales. Disciplined measurement in the social sciences must continue to press forward building on new techniques and data sources arising from analytical and computational sociology focused on specific problems (Adams, 2001; Hatt, 2008; Watts, 2017).

If a correlation between spatial movement and social capacity can be found through a formal methodology, we may be able to generalize the relative social resources (social capital) of a community based on more readily obtained physical movement data, something that would be highly useful given that GPS data is nearly ubiquitous. This possible linkage may fall in the range of the "necessarily approximate nature of useful results in data analysis" (Tukey, 1962), reflecting that data may be very useful even when we cannot fully explain all of the dynamics contributing to that data (Halevy et al., 2009). Unlike earlier work on space syntax and urban patterns, this work begins with social capital as a representation of relational structures and examines ways in which the spatial movements of humans may be related to that particular representation. The focus of this paper explores the possible use of mathematical, statistical and machine learning tools to understand these potential relations.

In physics, the pressure or temperature of a gas can be obtained without knowing the state or nature of atoms or molecules that comprise the gas. As an example, astrophysics advanced for a considerable time through partial understanding about star composition based on observation – "opening a problem to investigation" – and inference despite opposition from formalists who wished to advance only on what could be determined with mathematical certainty (Stanley, 2007, p. 58). The tension between venturing a possibility and proceeding only on certainty was held in fruitful tension in that case and a similar effort is being made in this paper where a proposed method, using spatial movement as a proxy for social capital levels, is applied to the design, collection, and analysis of a complex data set. The measurement of social capital through an aggregated value does not presume to capture exhaustive knowledge of all the interacting dynamics involved in social capital, income, or geographic movement but considers the possibility that the aggregation of those elemental interactions produce measurable and meaningful variations. These variations arise out of the perennial and dynamic social hierarchies that human society has manifested across time and space (Adams, 2001; Zhou et al., 2005) and which also appears in large social data sets that evidence scaled structures within them (Zhou et al., 2005). Social capital has a long history (Aristotle, 2012; Tocqueville, 2001) with more recent and rapidly growing formal interest (Franke, 2005; Friesen, 2018a) and more gradual policy adoption (Ostrom & Ahn, 2003).

Given the complexity of social capital (J. S. Coleman, 1988; Hanifan, 1916; N. Lin et al., 2001), rapidly developing social data sources and machine learning techniques that allow for useful pattern detection in large data complex data sets such as in the current study (Keuschnigg et al., 2018) make ongoing research for novel signals within complex social data an important pursuit. The working definition of social capital (Chapter 1 Section 1.5) refers to the density and quality of formal and informal human relational networks. As noted earlier (Chapter 1 Sections 1.2 – 1.4), the primary research question, central hypothesis, and secondary hypothesis are:

Primary research question: *What is the relationship between how a statistically significant group of people move around their neighbourhoods, reflected in GPS spatial data, and social capital levels as established by the SCGSS baseline measurements?*

Central hypothesis is that social capital levels decrease with decreasing spatial movement. This hypothesis is derived from the logic that an individual who has never left their home will have fewer social contacts (social networks), lower levels of community institutional interaction (trust), and as a result, lower levels of social capital than an individual who is very active through work, school, shopping, recreation, and other human activities.

The **secondary hypothesis** explores the relationship between income levels and social capital. It has been noted that social resource levels can be low even among people with higher incomes, and social isolation appears to be *a phenomena that does not correspond directly with income* (Lindström & Lindström, 2006; Macinko & Starfield, 2001; Pearce & Davey Smith, 2003). Social capital studies are very often concerned with socio-economic dynamics, often presuming that social capital levels follow income patterns. The resolution of the survey instrument, the data collection strategy within Census Tract scales, and the exploratory nature of this paper makes the inclusion of an income hypothesis important.

4.2 Social Capital Data

Aggregating the variables related to social capital into a single value makes comparison with other variables such as income and the set of ten spatial variables easier with a linear model where high dimensionality (many variables) impairs the performance of the model. The total social capital variable is obtained by combining the first and second order synthetic social capital variables (social networks and trust) as described visually (Figure 4-1).

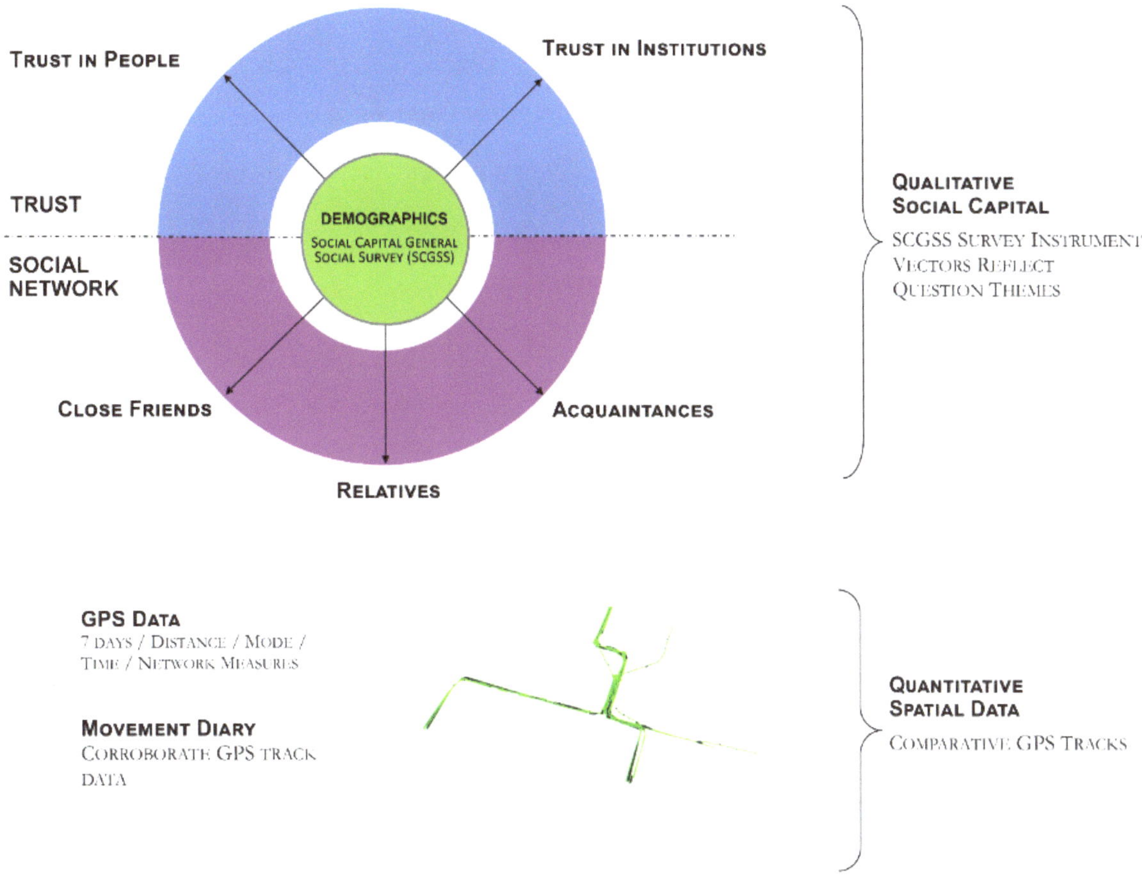

Figure 4-1 - Diagram of SCGSS social capital survey instrument question themes (demographic, social network and trust) along with an image of a GPS data track that comprises the spatial measures.

For synthetic social capital variable analysis, the following data frame changes have been made prior to the analytic calculations. For example, if *number of close friends* is included in the social network topography synthetic value, the *number of close friends* is counted twice and would skew model results. Therefore, wherever a synthetic variable is used, the sub-variables that contributed to it are excluded. This is to ensure that collinearity in the independent variables is taken into consideration prior to modeling and based on domain level knowledge (further statistical tests of collinearity are carried out during model refinements). For example, the ***Total Social Capital Value (all.socap)*** variable represents a synthesis of 22 original variables while the ***Social Network Synthetic Element (sns.socap)*** represents a synthesis of 9 original variables and so on. This gives us three distinct sub-data frames.

4.3 Spatial Data

Spatial analysis is based on 10 spatial measures derived from an extensive review of possible ArcGIS tools and processes including network analyst and associated approaches (see Chapter 3 and Table 3-2 for details). These measures are not the only possible candidates but reflect a representative mix of distinct spatial category types including distance traveled, area, compactness of points, directional tendencies, speed, statistical distribution, and probability of paths being random.

4.4 Methods for Comparing Spatial and Social Capital Data

Comparative data dynamics include an aggregated, synthetic form of trust levels and social network features alongside the spatial measures. Linear regression, random forest and principle component analysis data exploration approaches are used with carefully structured data gathering approaches (Breiman, 2001b; Pflieger, 2008).

Although individual level result reporting is outside of the ethics agreement used when collecting the data[8], the highest and lowest aggregate social capital level individuals do show clear income variance – high social capital high income and vice-a-versa. This is one of the clues discussed toward the end of the paper suggesting that the median income category and standard deviation above and below that are less distinct regarding income/social capital relationships. The exploratory search for a spatially inferred social capital proxy may need to be developed at the extremities and moved toward the middle rather than the approach taken here of beginning in the middle and moving outward.

If the effect of a social support or intervention could be seen both more clearly and sooner, it follows that investments of time, money and human resources could be more effectively directed, leading to approaches that strengthen rather than erode our social fabric (Forbes & Wainwright, 2001; Mario L. Small & Adler, 2019). The behavioural nature of the GPS spatial data gathered, a form of data that does not depend on answering perception or recall questions, is a promising source of insight and could lead

[8] I originally collected this data while at the University of Waterloo and their Human Ethics Review policies informed the agreements made with participants. The relevant passages from the "Social Imaging Study – Letter of Consent" signed by participants includes clarification that "...data will be kept anonymous in the findings...your data will be reported only in aggregate, and you will not be identified in any reports of the data."

to a more reliable form of social infrastructure mapping (T. J. Allen & Henn, 2006; Friesen, 2013b; Shoval & Isaacson, 2009).

The analysis which follows begins with linear regression to examine variables directly related to the central and secondary hypotheses and if such a relationship is found, the degree and nature of that significance. Choosing the variables directly related to the hypotheses reduces the dimensionality of the large 51x97 data frame. Following that, random forest modeling is used to examine if either the linear regression insights are supported or other patterns in the data not directly related to the hypotheses exist that merit further examination. Finally, principal component analysis is used on the full data frame to add additional support or refutation to the linear regression and random forest models. Random forest and principal component analysis are suitable for high dimensional data such as that suggested by number of variables in the data (Bivand, 2013; Breiman, 2001b; Hoaglin, 2003). Principal component analysis is a very old, well-established method of examining high-dimensional data for possible relationships among the variables. The current study is exploratory and experimental, making a wide range of methods and an additional corroborating data set useful. It is important to avoid both over- or under-attending to possible patterns in the data. This is consistent with a scientific investigation that is responsive to indications and partial answers (Tukey, 1962, p. 60).

The challenge of comparing spatial and social datasets is compounded by the complex nature of the variables and their interactions. The search for meaningful signals at an exploratory stage must be done carefully but there is a growing understanding that digital data generation through mobile devices and embedded devices leading to new forms of human behavioural data collection based (Deng et al., 2018; Goodchild, 2007). Advances in machine learning (automated or semi-automated high through-put statistical learning) have led to many new approaches and techniques in both supervised and unsupervised learning (Hastie et al., 2011, pp. 485–486). In this paper, the goal is to formally analyze social and spatial data based arising from urban movements and survey data focused on trust and social networks. Each of these models and their refinements (along with additional details) are available as R Markdown packages for reproducibility(Leek & Peng, 2015) and are discussed in more detail below.

With three different analytic approaches being used to compare social and spatial data, each with unique coding and refinement requirements, a control data set was introduced to verify that the *function of the various methods* was performing as expected, including the details of the custom coding for each. Evaluation data consisting of training and test data collected by the study was used within the different methods where applicable. The ***mtcars*** dataset was used to test the method approaches. It was published in 1974 by Motor Trend US magazine and includes mileage values, weight, horsepower, quarter mile performance and other details (11 variables) for 32 different models of cars manufactured from 1973-1974. The documentation and datasets in R provide these details. This data set has been analyzed almost continuously in a great many ways through the R repository and other. The results of analyzing ***mtcars*** using the linear, random forest and principal component analysis methods and coding in this paper are congruent with expected, published results (Henderson & Velleman, 1981). Features such as clear relationships between horsepower, weight of the vehicle, engine displacement and the fuel efficiency of a particular vehicle show up clear in the coded linear model, the random forest and principal component analysis coding. This suggests that the difficulty in extracting patterns in the data are the result of the data rather the methods employed.

4.4.1 Linear Model

In R, the linear model *lm()* uses bivariate analysis. Linear regression models work best with low numbers of variables, so the analysis is structured in a more fine-grained way to achieve this. One characteristic of this need for bivariate regression is that there are more steps and sub-steps in the way that the data is organized. Normalization of variables is also required for accurate analysis of a linear model.

Variables that are directly related to the central and secondary hypotheses have been examined using a linear model to test for meaningful variable interactions. The linear model function in R *lm()*, uses least squares linear regression structured so that, in its simplest form, a single independent variable is used to predict the outcome of a single dependent variable (Chambers & Hastie, 1991). The default correlation used is Pearson (*method="Pearson"*) unless otherwise specified in the script (e.g. *method="Spearman"* or *"Kendall"*). Training and test sets are used to test the validity of the model using data from the same set. The coding in R was set up so that the train/test function is drawn from the variables being examined. This was tested using the full data set and a second data set that contained only the two variables being examined. The summary statistics for both tests were identical.

4.4.2 Random Forest Model

The random forest (RFM) machine learning approach *randomForest( )* uses a different data analysis strategy. Rather than looking for linear associations between two variables, data is analyzed by generating a significant number of decision trees based on which data elements related best with each other at each branch and then tabulating those results to create a probability structure for interpreting new data. This is a form of unsupervised, ensemble learning that uses the aggregation of trees to identify patterns and signal those patterns potential significance (Neville, 1999) and it is not sensitive to multicollinearity given that at each variable split, only one is chosen so only one of the two collinear variables is selected. Random forest algorithmic approaches also work very well with high dimensional data and are robust with regard to noise, missing values, and non-normal data distributions (Breiman, 2001a; Chambers & Hastie, 1991; Khabsa et al., 2016; Moguerza & Muñoz, 2006; Morgan & Sonquist, 1963; Neville, 1999).

RFM does not begin with a model that the data is explored against as is the case with classical regression analysis. Instead, this ensemble method uses an algorithm that produces a large collection of de-correlated trees (each is generated independently of the others) and the predictive performance of this random forest of trees compared and tallied (S. Kuhn et al., 2008). In particular, the RFM may be understood as another tool that enables us to make more educated guesses about what is meaningful in a given set of interacting variables through listing variables by order of importance or influence on the model.

The decision trees generated by RFM may be structured by classification (finite number range) or by regression (continuous values as in real numbers) (Salzberg, 1994). The dependent variable type determines which approach is taken. Classification uses a voting approach for which trees fit best while regression uses the mean of the output of each tree In what follows, the statistical software R 3.2.5 (R Development Core Team, 2016) with the Random Forest package is used to undertake a combination of classification and regression machine learning analyses. Analysis of the data is undertaken in R to make use of a TEST set comprised of 20% of the actual data set randomly selected and a TRAINING set comprised of 80% of the data.

The flow of the algorithm begins with the original data and produces a collection of decision trees that have been sorted, evaluated, and their characteristics compared with the TEST set recorded (Figure 4-2).

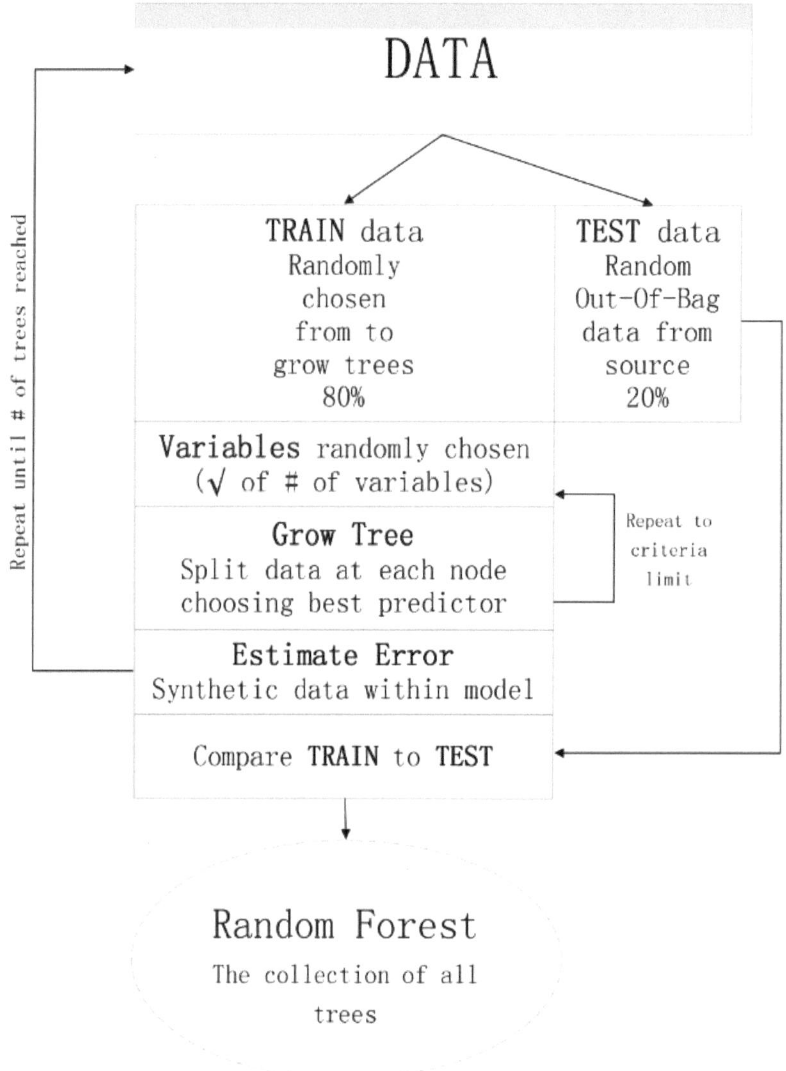

Figure 4-2 - Random Forest algorithm logic. For variable selection, it should be noted that the rfm() model in R determines the number of variables chosen by using the square root of total variables for classification random forests (i.e. dependent variable is a factor) and the number of total variables divided by 3 for regression random forests (i.e. dependent variable is continuous). The number of variables can be specified manually.

4.4.3 Principal Component Analysis

Principal component analysis(PCA) has been around since 1901 (Salzberg, 1994) and is an analytic method for examining data that helps us see the most important (highest information) structures (Jolliffe & Morgan, 1992, p. 69). PCA calculates a hyperplane – a way of cutting the 3 dimensional plot of data – that best differentiates structures in the data. This approach is suited to exploration of high dimensional data, is not affected by collinearity, and is useful for identifying the degree to which given variables contribute to those distinctions (Bibby et al., 1981; Venables & Ripley, 2003). PCA may be based on the correlation matrix of the variables or on singular value decomposition (see R documentation "prcomp" and "princomp as well as Mellinger (1987)). The later PCA method is used in this analysis. Given these characteristics, PCA is reasonably well suited for use as a variable reduction method (Suhr, 2005, p. 1) since it shows us the axes where variation among variables is greatest and which variables, consequently, can be set aside . In this paper it provides corroboration for variable interactions between the social capital and spatial datasets being examined including the synthetic dependent variable ***all.socap*** and the ten independent spatial variables. Proportions of change among variables are calculated using both table formats and vector diagrams as outputs. These summaries reveal which independent variables share common contributions to changes in the dependent variable (Bibby et al., 1981; Mandel, 1972; Venables & Ripley, 2003; Young et al., 1978).

Variances can be plotted by component where many variables are present using a scree plot (Cattell, 1966) to determine visually where the proportion of variance is not significant – generally for variance values less than 1 and reflected by a flattening of the plotted curve (Yeomans & Golder, 1982). Accumulated variance can also be used to determine where the component cut-off should occur – e.g. include all components that are required for 90% variance (Costello & Osborne, 2005). The higher these values with the fewest number of components, the more likely it is that the structures in the data can be seen. The goal of the PCA method used here is to interpret the results against what has been observed in both the linear models and random forest models above through identification of structural similarities and differences.

4.5 Results

The results of the three foregoing analytic approaches are summarized below.

4.5.1 Summary of Linear Model Results

A boxplot and histogram of ***all.socap*** (the total social capital score) shows that the data reflects a normal distribution of values (Figure 4-3).

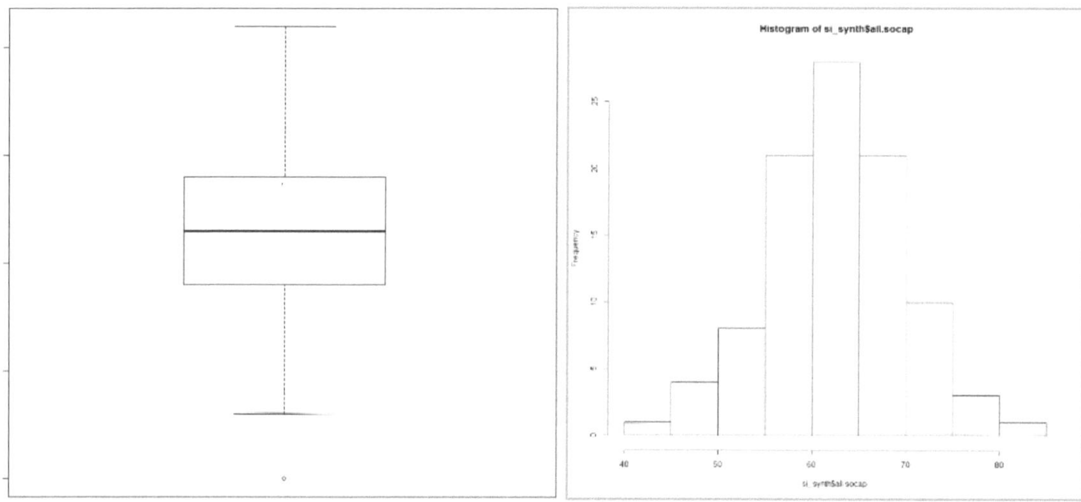

Figure 4-3: Total synthetic social capital value (all.socap) distribution across all 97 participants.

This is important for subsequent analysis as the linear model function in R assumes that the dependent variable – in this case ***all.socap*** – is normally distributed and does not require further transformation (e.g. BoxCox). The boxplot reflects that there are outliers to consider when tuning the model. Density plots and histograms were used on all variables to check for normal distribution and with all variables except ***all.socap***, ***nearest neighbour z-score***, ***total.distance***, ***ellips.rotat***, ***standard.speed***, a log(10) transformation was applied to bring values closer to a normal distribution. Data was partitioned into an 80/20 Train/Test set for estimating model accuracy after tuning. A check for the linearity of errors was made using a plotting function that produces a model fit line and a line fitted to the distribution of residuals. These reflected sufficient congruity to proceed with further model refinement. After running the initial model, outliers were identified using Cook's distance and removed until the R^2 model fit began to decline. With extreme values removed from the model, the correlation of independent variables (eg. having "age" and "years alive" in a model would lead collinearity) was checked using a variance inflation factor (VIF) function. Five independent variables had VIF values above 5 and were eliminated. The linear models ***LM1***, ***LM2***, and ***LM3*** results show that after taking these model tuning steps with the balance of variables, running models on new test data, and summarizing results, no commonly significant statistical results ($p>0.05$) were obtained (Table 4-1).

Initial Models – normalized variables with all independent variables included in model						
Linear Model	Central Hypothesis Social Capital and Spatial Patterns	Dependent Variable	Independent Variables	Correlation TRAIN (R^2)	Correlation TEST (R^2)	Adjusted R^2
LM 1		all.socap	$ total.distance	6.2%	2.6%	0.2%
LM 2		sns.socap	$ standis.radius	5.3%	2.6%	2.2%
LM 3		ts.socap	$ dif.memed	9.6%	0.09%	1.5%
			$ ellips.rotat			
			$ ellips.xyrat			
			$ nn.obmedist			
			$ nn.ratio			
			$ nn.zscore			
			$ allpointsarea.rectangle			
			$ standard.speed			
	Secondary Hypothesis Social Capital and Income					
LM 4		all.socap	income	4.8%	0.49%	3.3%
LM 5		all.socap (Delta)	income (Delta)	0.09%	44.9%	-4.5%
LM 6		all.socap (Rosedale)	income (Rosedale)	4.8%	28.9%	-2.1%
LM 7		all.socap (McQueston)	income (McQueston)	0.36%	7.3%	-3.3%
LM 8		all.socap	neighbourhood	6.25%	4.0%	3.4%

Tuned Models – sensitivity to outliers(Cook's distance), variable inflation factor (VIF), multicollinearity, TEST holdout data		
Linear Model	Final Model Variables	p-value
Central Hypothesis Social Capital and Spatial Patterns		
LM1	total.distance	0.129
	dif.memed	0.966
	ellips.rotat	0.480
	ellips.xyrat	0.540
LM2	ellips.rotat	0.439
	ellips.xyrat	0.069*
LM3	ellips.rotat	0.156
	ellips.xyrat	0.211
	dif.memed	0.466
Secondary Hypothesis Social Capital and Income		
LM4	income	0.067*
LM5	income(Delta)	0.890
LM6	income(Rosedale)	0.423
LM7	income(McQueston)	0.534
LM8	all.socap ~ neighbourhood(McQueston)	0.051*
	all.socap ~ neighbourhood(Rosedale)	0.110

Table 4-1 Initial and tuned model results summary.

Based solely on the numerical relationships in the data, the most significant correlation with a TRAIN dataset was the *trust* component of social capital where the final significant variables were ***ellips.rotat***, ***ellips.xyrat***, and ***dif.memed***. None of these final model variables showed notable p-value scores suggesting that the only value that we may be able to derive from the results is a clue about which spatial elements are relatively important vis-a-vis the others spatial measures. To the extent that we follow this weak signal, it appears that an increase in the ***ellips.xyrat*** (which means the territory covered by an individual is more circular around the mean spatial centre rather than being elongated directionally) is reflected in an increased aggregate trust score (***ts.socap***) as a component of social capital. The ellipsis measure characteristics (increases in social capital with more circular territory and greater orientation toward an East/West travel pattern) show up in ***LM1*** (single synthetic social capital score), ***LM2*** (social network component of social capital) and ***LM3*** (trust component of social capital). This result is likely trivial and is not a meaningful result in terms of generating deeper insight about the relationship of social capital with spatial movement.

Another weak suggestion of measurement value comes from ***all.socap*** in relation to ***total.distance***. Although the p-value is not very significant (0.129) relative to other measures, it is by far the most significant variable and suggests there may be a positive correlation between increased total social capital scores and increased total distance travelled. This weakly supports the central hypothesis that more travel and increased social capital are positively correlated. It is also the most easily interpreted as a behaviour with greater movement understood readily as related to a larger range area.

The null form of the secondary hypothesis asserts that a relationship between income and social capital does not exist. Both the individual income (***income***) and social capital level (***all.socap***) for each of the 97 instances were examined and also in aggregate form for the three income-differentiated Census Tracts via models ***LM4 – LM8***.

For the secondary hypothesis, a significant p-value score of 0.067 weakly suggests there may be an increase in total social capital with increases in income. The most significant Census Tract segment was the McQueston neighbourhood where the p-value of income and total social capital was > 0.051. A possible interpretation is that in the lower income Census Tract, an income / social capital correlation begins to be seen which may be obscured when averaging takes place across the wider, more income distributed data set. This result is mitigated to some extent by an inconclusive correlation between the TRAIN and TEST data across all income-driven models. The small sample size of 97 makes the TRAIN and TEST approaches less valuable.

Given the exploratory nature of this research design and data set, the random forest machine learning model will be used next to explore other possible relationships among the variables, including comparison with relative variable importance noted in the linear model.

4.5.2 Summary of Random Forest Model Results

The following section analyzes the ***synthetic variables*** by using RFM in R. Each participant has both the 25 individual social capital variables and the synthetic measures. Variables used to calculate synthetic variables are held out in the subsequent analysis to avoid direct collinearity. Both individual and synthetic values are examined using the RFM model algorithm for each of the hypotheses reflected in ***RFM1-RFM8***. These are arranged in response to the central and secondary hypotheses and follow the Linear Model analytic structure. Data was partitioned into an 80/20 Train/Test set. A decision tree model was developed initially to identify which variables the data split on for the various models and hypothesis. The RFM was tuned by number of trees with an optimum number identified by using error rate vs. number of trees plot – typically 500 trees were used. The primary interest in using RFM is to identify the relative importance of variables. One of the important advantages of RFM is identification of information patterns in noisy, high dimensional data. These characteristics provide a counterpoint to the foregoing linear model where sensitivity to normalization, collinearity and the number of variables are all complications that model tuning must account for.

The central hypothesis asserts greater movement will lead to higher social capital and the most important spatial feature when the total social capital score (***all.socap***) was run as the dependent variable for the area of the territory(***allpointsarea.rectangle***) encompassed by an individual's movements over a seven day period. The earlier discussion about individual track characteristics and the specific dynamics of commuting, urban form, and localized activities would apply here if that level of granularity was in mind. The study, however, is focused on the aggregated neighbourhood (Census Tract) scale pattern where individual features (constraining to street paths, modal choice, work and leisure patterns) are not singled out. At an individual level, variances such as Euclidean vs. Manhattan distance, walkability, and related dynamics would be important. As noted in the table of results, 5.51% of the variance in total social capital (***all.socap***) is explained by the total area measure (***allpointsarea.rectangle)*** (Table 4-2).

Random Forest Model	CENTRAL HYPOTHESIS Social Capital and Spatial Patterns	Dependent Variable	Independent Variables	Top Variable	%IncMSE
RFM 1		all.socap	$ total.distance	allpointsarea.rectangle	5.61%
RFM 2		sns.socap	$ standis.radius	nn.zscore	7.33%
RFM 3		ts.socap	$ dif.memed	nn.obmedist	5.32%
			$ ellips.rotat		
			$ ellips.xyrat		
			$ nn.obmedist		
			$ nn.ratio		
			$ nn.zscore		
			$ allpointsarea.rectangle		
			$ standard.speed		
SECONDARY HYPOTHESIS Social Capital and Income					
RFM 4		income	$ close.number	acquaint.contact	4.51%
RFM 5		income(Delta)	$ close.contact	trust.locbiz	3.59%
RFM 6		income(Rosedale)	$ close.freq	trust.fam	3.80%
RFM 7		income(McQueston)	$ close.walk	acquaint.number	3.44%
RFM 8		neighbourhood	$ aquaint.number	trust.locrelig	6.49%
			$ acquaint.contact		
			$ acquaint.freq		
			$ acquaint.walk		
			$ relative.number		
			$ relative.contact		
			$ relative.freq		
			$ relative.walk		
			$ trust.people		
			$ trust.fam		
			$ trust.neigh		
			$ trust.classmate		
			$ trust.strange		
			$ trust.police		
			$ trust.pubschool		
			$ trust.city		
			$ trust.bigbiz		
			$ trust.locbiz		
			$ trust.locrelig		
			$ trust.healthcare		
			$ trust.welfare		

Table 4-2 Results of random forest analysis for central and secondary hypotheses.

The social network component of the social capital measures indicate the degree of randomness present in a spatial movement pattern (***nn.zscore***) is the most important variable explaining 7.33% of the variation while the trust component (***ts.socap***) revealed that compactness of a spatial network (***nn.obmedist*** reflecting shorter average distances between readings) is most important explaining 5.32% of the trust score variance.

The secondary hypothesis asserts social capital increases are not correlated with income levels. The most important variable with regard to ***income*** and the 25 non-aggregated social capital measures was the mode of acquaintance contact (***acquaint.contact***, e.g. face-to-face, digital, phone) with 4.51% of the income variance explained by that variable. When the income modeling was segmented by neighbourhood, the most important variable for Delta (upper income) was the degree of trust in local businesses (***trust.locbiz*** with 3.59% of variance), for Rosedale (middle income) it was degree of trust in family members (***trust.fam*** with 3.8%) and for McQueston (lower income) it was the number of acquaintances (***acquaint.number*** with 3.44%). This diffusion of results suggests support for the secondary hypothesis that income and social capital are not correlated. Within the parameters and assumptions of this study, it does not appear that social capital levels follow income (either positively or negatively) whether all 97 participants are examined as a group or are segmented by income differentiated Census Tracts. When the dependent variable is neighbourhood and the independent variables are the individual 25 social capital elements are modeled using a logistic regression, the most significant variable is compactness of the spatial network where in RFM2 (***nn.zscore*** with 7.33%) followed by trust in local religious groups (***trust.locrelig***) contributing 6.49% of the neighbourhood variable explanation.

4.5.3 Summary of Principal Component Analysis Results

When interpreting the results of PCA, it is important to keep in mind that the matrix of factors produced from the variables are not formally interpretable (Mellinger, 1987; Suhr, 2005, p. 11). This means that we cannot draw a linear conclusion about the increase in one variable correlating to an increase in another formally as in a linear regression. The value of PCA is understanding relative variance across all the variables examined – which variables contribute most to the variance in the data, such that we know which variables could be used in regression analysis where high dimensionality would be a problem. Domain knowledge of the data is important when choosing which component to evaluate and to be cautious since "when principal components are used in regression or discrimination it is not necessarily always the components with the largest variance which provide the best prediction" (Jolliffe & Morgan, 1992, p. 92). It may be that PCA analysis reveals that there aren't any variables that contribute significantly to variance in the data. In our case, the ***LM1*** model shows that a variety of spatial variables are highly correlated. It was expected that this would also be picked up in the PCA analysis and this occurred with ***total.distance*** and ***allpointsarea.rectangle*** contributing to Principle Component 1 (PC1) (Table 4-3). This means that there are not clearly important variables in the total social capital comparison with spatial variables. In turn this would mean, at least in part, that we cannot feasibly reduce the number of variables used in a linear model. Our earlier results from the linear model reflected this. A more useful result would be to see fewer variables with much higher proportional variance for PC1 and PC2. Such is not the case.

<table>
<tr><th>Principal Component Analysis Model</th><th>CENTRAL HYPOTHESIS Social Capital and Spatial Patterns</th><th>Dependent Variable</th><th>Independent Variables</th><th>PC1 proportional variance</th><th>PC1 Key Variables</th><th>PC2 proportional variance</th></tr>
<tr><td>PCA 1</td><td></td><td>all.socap</td><td rowspan="3">$ total.distance
$ standis.radius
$ dif.memed
$ ellips.rotat
$ ellips.xyrat
$ nn.obmedist
$ nn.ratio
$ nn.zscore
$ allpointsarea.rectangle
$ standard.speed</td><td>39.84 %</td><td>total.distance
standis.radius
allpointsarea.rectangle
standard.speed</td><td>13.60%</td></tr>
<tr><td>PCA 2</td><td></td><td>sns.socap</td><td>39.72%</td><td>standis.radius
allpointsarea.rectangle
standard.speed</td><td>13.50%</td></tr>
<tr><td>PCA 3</td><td></td><td>ts.socap</td><td>39.78%</td><td>total.distance
standis.radius
allpointsarea.rectangle
standard.speed</td><td>13.85%</td></tr>
<tr><td></td><td colspan="6">SECONDARY HYPOTHESIS
Social Capital and Income</td></tr>
<tr><td>PCA 4</td><td></td><td>income</td><td rowspan="5">all.socap</td><td>57.37%</td><td rowspan="5">[trivial]</td><td>42.63%</td></tr>
<tr><td>PCA 5</td><td></td><td>income(Delta)</td><td>60.79%</td><td>39.21%</td></tr>
<tr><td>PCA 6</td><td></td><td>income(Rosedale)</td><td>60.92%</td><td>39.08%</td></tr>
<tr><td>PCA 7</td><td></td><td>income(McQueston)</td><td>51.85%</td><td>48.15%</td></tr>
<tr><td>PCA 8</td><td></td><td>neighbourhood</td><td>58.99%</td><td>41.01%</td></tr>
</table>

Table 4-3 Principal component analysis results summarized.

The graphical output of the table values provides another view of the PCA result and confirms the clustering that we see numerically. In all cases, there are not reliably clear structural differences whether with social capital measures (Figure *4-4*), social network related social capital variables (Figure 4-5) or with the trust related social capital variables (Figure 4-6).

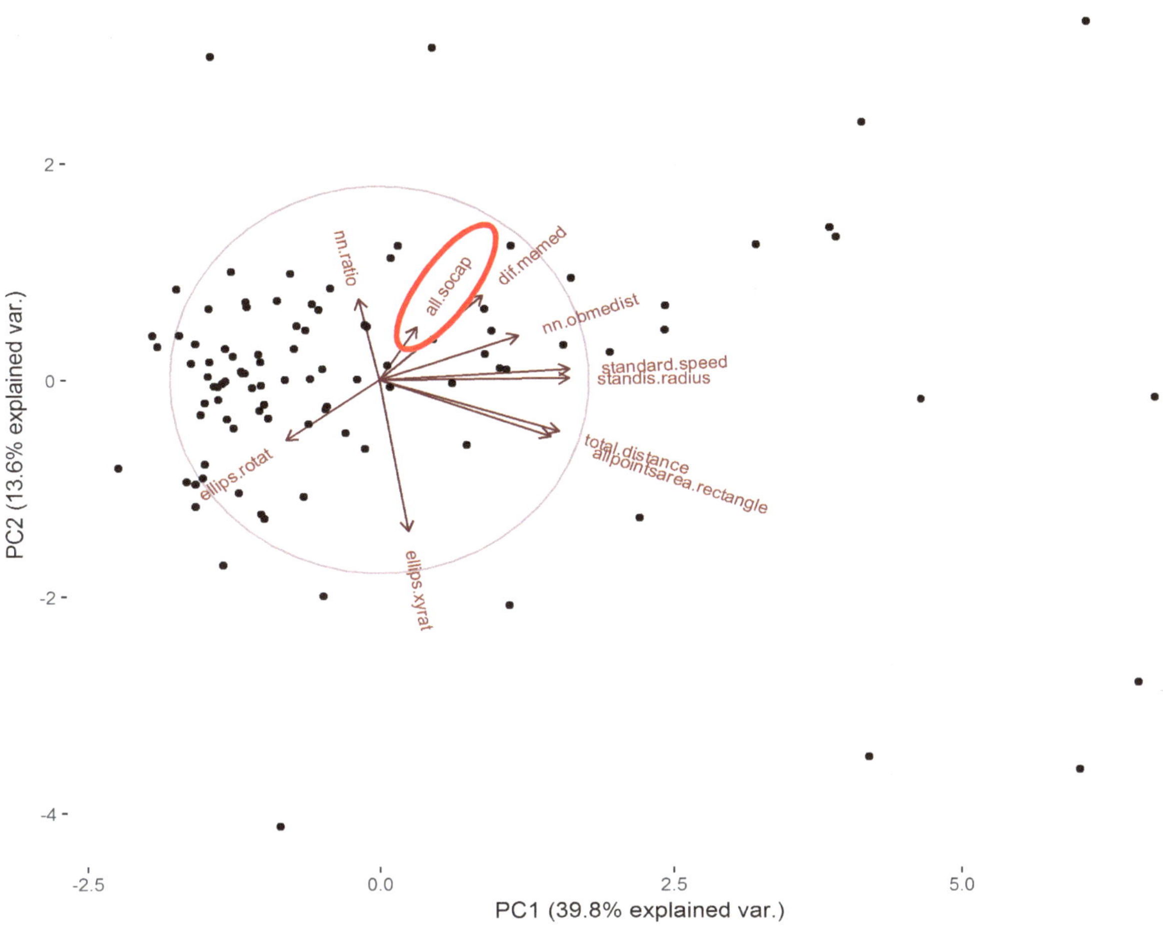

Figure 4-4 - Principal component analysis for **all.socap** (PCA1 model - total social capital) variables with the circle indicating the spread of 50% of the data points.

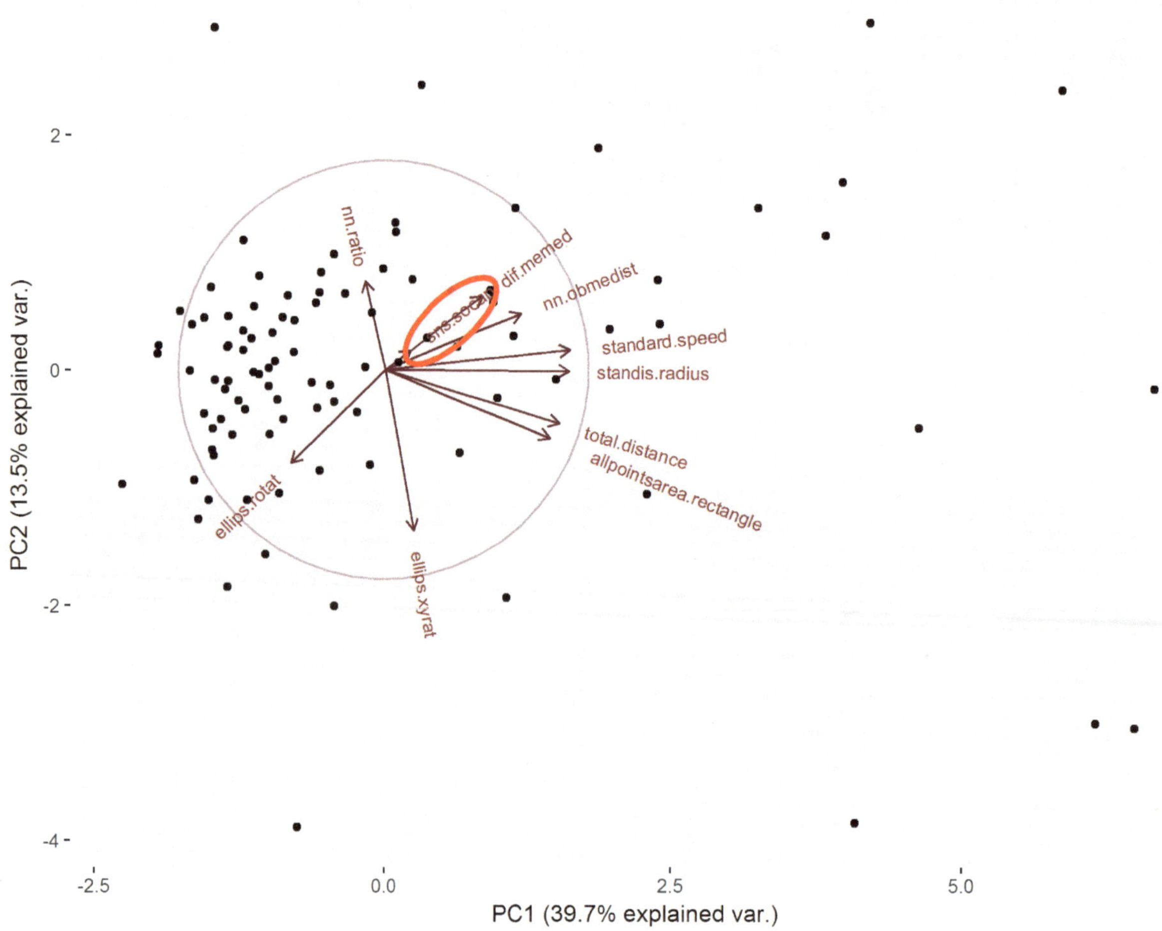

Figure 4-5: Principal component analysis for sns.socap (PCA2 model - social network related social capital) variables with the circle indicating the spread of 50% of the data points.

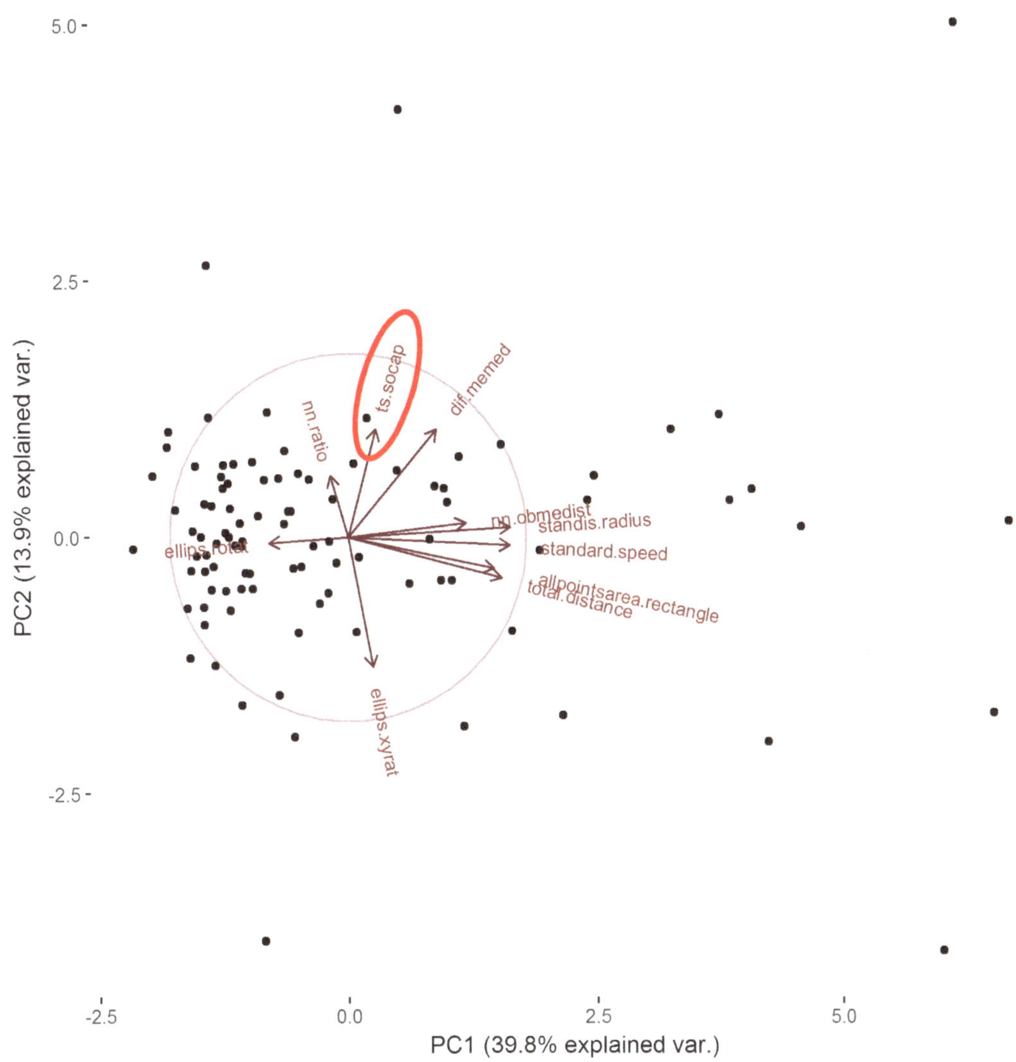

Figure 4-6: Principal component analysis for **ts.socap** (PCA3 model - trust related social capital) variables with the circle indicating the spread of 50% of the data points.

4.5.4 Comparison of Linear, Random Forest and Principal Component

Before undertaking a comparison of our various model results, it is important to introduce a corroboration employed as a broad test of validity. As a form of methodological validation, the LM, RFM and PCA approaches used in this paper, including coding structures, are applied to the well-established ***mtcars*** data set in R where results and patterns have been independently identified (Henderson & Velleman, 1981). This comparative review provides increased confidence in the application of these approaches to the current, much more experimental data sets being explored and answers the question of whether the methods used would have revealed important structures in the data if those structures are in fact in the data collected.

The following summary is a comparative examination of the results of the linear model, the random forest model and the principal component analyses. The first part of the summary concerns the central hypothesis which argues that increased travel leads to increased social capital levels and the secondary hypothesis that social capital dynamics do not follow income levels directly. One of the central goals of this research project was to understand how spatial habits and income relate to social capital levels. The 97 participants in the study provided a valuable socio-spatial data set for initial examination. As the preceding results suggest, the social and spatial patterns of people in cities represent a highly complex, richly interacting set of phenomena. In general, all three forms of analysis suggest that there isn't a single, dominant driver that relates social capital levels to spatial movement patterns (Table 4-4).

Significance Table for Linear, Random Forest and Principal Components		
	Independent Variable	p-value
LM1 – all.socap	total.distance	0.129
LM2 – all.socap	ellips.xyrat	0.069
LM4 – all.socap	income	0.067
LM8 – all.socap	McQueston income	0.051
	Top Variable	**%IncMSE**
RFM1	allpointsarea.rectangle	5.61%
RFM2	nn.zscore	7.33%
RFM3	nn.obmedist	5.32%
RFM4	acquaint.contact	4.51%
RFM5	trust.locbiz	3.59%
RFM6	trust.fam	3.80%
RFM7	acquaint.number	3.44%
RFM8	trust.locrelig	6.49%
	Component 1 Proportional Variance	**Key Variables**
PCA1	39.84%	total.distance, standis.radius, allpointsarea.rectangle, standard.speed
PCA2	39.72%	standis.radius, allpointsarea.rectangle, standard.speed
PCA3	39.78%	total.distance, standis.radius, allpointsarea.rectangle, standard.speed

Table 4-4 *Summary and comparison of linear model, random forest and principal component analyses.*

One of the common variables of significance across all three models was ***total.distance***. In the linear model the ***all.socap*** run against ***total.distance*** had a p-value of 0.129 which is not within the 0.05 p-value significance range but compared with the significance level of other variables, is notable. The random forest model did not rank ***total.distance*** as a top variable but ***allpointsarea.rectangle*** did show up, on a ranking basis, as a top contributor to model error if it was not included (5.61 %IncMSE – increase in mean square error). The total distance travelled by a participant is expected to have some relation to total area (***allpointsarea.rectangle***) though high activity level in a smaller overall area (traversing same local route repeatedly) is possible. The random forest model, showed an almost unique set of top five variables across the three Census Tracts. This suggests that income and social capital interactions are complex. Additional analysis could be performed by using the random forest method as a starting point and then running more direct linear models on each of the top variables. It is unlikely that this would yield improved results. Some commonality appears when ***neighbourhood*** as dependent variable and the 25 social capital variables as the independent variables are run. In that case, an order of importance emerges with ***trust.locrelig***, ***trust.neigh***, ***close.freq***, ***close.number***, and ***acquaint.number*** showing up as the top five variables.

The principal component analysis showed included ***total.distance*** and ***allpointsarea.rectangle*** in the same cluster and perpendicular to ***all.socap*** (Figure 5). As noted in the preceding paragraph, this is generally expected. For the linear model (***LM2***) the ***ellips.xyrat*** had a p-value significance of 0.069 suggesting that an elongated spatial pattern had some correlation with social network characteristics. Ellipsis characteristics were much more notable in the linear models but did not show up at all in the random forest models or in any significant way in the principal component analysis. This may also be owing to procedural issues but could be a result of there being no relationship between social capital and the median summary of points in a set of GPS data records.

The relationship between income and social capital levels explored through the secondary hypothesis suggest, with the linear model, that there is a correlation between ***all.socap*** and ***income*** with a p-value of 0.067. The ***income*** variable in the random forest model reflected a relative ranking of importance for ***acquaint.contact*** as a top ranked contributing variable split. Among the balance of other random forest models, however, each model had a different top ranked variables which precludes identification of a clear income to social capital relationship. In each case of top variables, their contribution did not have much significance in terms of the increase in mean square error (%IncMSE). PCA results for the ***PCA4-***

PCA8 models show graphically that with two principle components very little distinguishing structure among the data points occurs – the first and second components are almost evenly split in their structural difference so that distinguishing the dependent and independent variable interactions is not meaningful.

4.6 Discussion

The goal in designing this research was to explore potential correlations within the complex matrix of social and spatial dynamics in East Hamilton. Labour intensive interviews and data gathering (outside of online social media sources) make social network research difficult except for small studies. However, with new forms of observation driven data collection, such as GPS data logging, we may be able to improve empirical knowledge of social capital levels with neighbourhood resolution across a city or region and thus direct limited city planning public resources more effectively (Aaltonon et al., 2005; Bailey et al., 2019). Improved measurement of social capital and emerging approaches to these types of latent social structures (Borsboom, 2008; J. S. Coleman, 1986; Costello & Osborne, 2005; Evans & Aceves, 2016; Grimmer, 2015) means greater knowledge of the social impact of planning, social support, and other forms of urban intervention (Sekar & Rai, 2018).

The foregoing results are suggestive rather than definitive. The ten spatial measures, modal analyses, and social capital elements do not exhaust the possibilities for finding social capital and spatial correlations. Additional methods, beyond the scope of this study, could include the analytical characteristics of the pathways taken by the participants (e.g. Route Choice Analysis (Dorninik Papinski & Scott, 2011)) as well as finer grained details about individual level behaviours along the GPS line recorded – where is work, where is home, what do the stopping densities represent beyond this, what is left out, and so on.

There exist, beyond the range of the central and secondary hypotheses in this paper, additional questions and possibilities that may be amenable to data mining algorithms that can find time/location/social interaction patterns from large datasets (Eagle, 2005; Sampson & Graif, 2009; Sampson & Raudenbush, 1999). The core research question concerning the relationship between how a statistically significant group of people move around their neighbourhoods, reflected in GPS spatial data, and social capital levels as established by the SCGSS baseline measurements, arises out of a need to improve social capital measurement at a local level. This scale has proven difficult to carry out laterally across a city due to the cost and complexity of capturing social capital dynamics. The use of GSS questions related so social capital proved useful in establishing social capital levels if a more efficient means of applying those questions to a broad urban population could be found. The spatial correlation aspect was intended to provide corroboration in the event that such an efficient means of direct measurement could not be devised and a secondary or corollary measure had to be used. Although definitive correlations that could be expanded generally were not identified, limited signals were identified that suggest this idea could be more fully examined and refined. Possessing such insight into social dynamics would significantly increase planners' ability to understand the needs of neighbourhoods and assess the impact – both positive and negative – of a wide range of planning developments. If the gap between changes in the physical design of a city and the social effect on citizens could be narrowed, planning would have the potential to better link design and accountability toward more human-centered and socially productive would be more possible than is now the case. Some of these design-oriented implications for social and spatial interactions will be examined more fully in the concluding chapter.

A range of possibilities to deepen understanding remain. There may be additional information that could be extracted about the relationships between the Spatial Data variables and the Demographic, Context and Social Capital variables than has been possible to explore here either practically or theoretically - e.g. the growing literature around space syntax and related neighborhood level design dynamics (Haas, 2008; Hillier, 2015; Major et al., 1997; Talen, 1999). If there is any consolation arising from the work undertaken in this early work, I hope it is that the intricate dynamics of the social infrastructure of our cities has been re-affirmed as a very challenging space theoretically and practically.

Complex social phenomena will never be easy to decipher. Social capital remains a new field (Ostrom & Ahn, 2003) without an agreed to definition suggesting significant scope for continued growth and enough ambiguity to lend it an adolescent stage of understanding. As noted earlier, there are tens of thousands of academic papers across dozens of disciplines reflecting a context of interest for social and spatial exploration. Growth in understanding complex social phenomena will take place through a combination of disciplined investigation that makes use of rigorous statistical and empirical processes along with exploratory connections and leaps that will prevent us from settling on locally optimized solutions that are far below a globalized optimum (Goldberg, 1989).

The hope is that if a more cost- and time-effective proxy measure for social capital can be found using ubiquitous, organic, dynamic spatial pattern data generated by billions of mobile devices and the Internet of Things, we will have been able to move from rudimentary visual maps of social infrastructure to much more extensive and accurate maps of the invisible landscape of social capital. This data will require suitable methods and adequate theoretical support in order to derive insight from it.

Demographic and context data was treated in aggregate form and converted into direct numeric values. This is suitable for exploratory pattern detection but a more fulsome explanation of the participant population by Census Tract would be needed to justify the form of commensuration being postulated between the statistical data and the human behavior being described (Wendy N. Espeland & Stevens, 1998).

Descriptive, exploratory and inferential modes of analyses are appropriate for the datasets that are being utilized but more concrete conclusions remain for future consideration. As such, the current paper provides one means of approaching the problems of social capital measurement by proxy using an originally generated dataset. It may well be that there are other data elements that are needed to support further extension of the social capital/spatial movement hypothesis.

The linear, random forest and principal component analysis approaches are three carefully chosen representatives among many dozens of other models and techniques (Chambers & Hastie, 1991; Hastie et al., 2011; R Development Core Team, 2016; Venables & Ripley, 2003). Although these approaches are well established, analytically distinct and have been used according to accepted practices, it could be that other modeling approaches could have yielded insights. It is also possible that the analysis, despite great care, contains errors in coding, data imputation, and distortions owing to improper normalization. Although great care was taken to avoid over-fitting, it is possible that tuning steps may also have introduced distortions into the analysis. Multiple modeling approaches and types were used to offset this possibility in substantive ways rather than relying on refining a single class of model. Full transparency related to data collection, cleaning, analysis and reproducibility also makes such errors visible to other analysists and researchers leading to corrections and improvements.

Care must be exercised in this process to avoid oversimplification of a complex phenomenon by means of a proxy measurement. It is not clear that there is such a measurement or that if found, it would be adequate to anything more than a very selective aspect of the phenomena. In the data collected and analyzed for this study, there is nothing to warrant causal inferences and there are very weak signals for possible correlations. The primary value coming from the study is identification of the relative value of variables and signaling of possible stronger correlations outside of the normal income distribution. Awareness that these boundaries exist is critical for future research. Despite the modest result, there are clues that the hypothesis may warrant further examination.

For example, when the highest and lowest total social capital values are examined at an individual level, an interesting dynamic is visible: social capital and distance traveled do increase and decrease together.[9] A definitive claim cannot be made on the bases of two cases but could be a clue for an adjusted data collection or analysis approach. The survey was also able to identify differences in trust, size of friendship, family and acquaintance networks, sense of belonging and other socially driven human dynamics. When the income patterns and social capital levels are plotted against the Census Tract income levels, there appears to be something unique happening at the very low end of the income scale suggesting that certain social capital features become evident under one set of circumstances and are submerged in another. There are known challenges for interpretation of data that arises across scales and structures (Anderson, 1972) but these challenges do not imply that insight is not possible.

The secondary hypothesis of this study, that social capital levels do not mirror income levels – i.e. a quantitative decrease income does not mean a corresponding decrease in social capital, was driven by other research suggesting that social poverty (low social capital) is present at the extremes of either poverty or wealth (Dixon & Frolova, 2011; Goldstein et al., 2009; Nisbet, 1953; White, 2010). In this paper, data collection focused on a central average and a standard deviation above and below the median adult income. While it is a limit of this research, the decision to focus on the median as a core reference was to minimize the effects of extremes at either low or high income. This may have eliminated important signaling outside of these value ranges. The low participant response rate (97 useable completions out of ~6000 attempted phone calls) for combined survey and GPS data collection means generalization beyond this study group is not possible. Simultaneous social and spatial data collection is difficult work but there is potential for insight and extension of methods.

---

[9] As noted earlier, ethics agreements dictate that individual results cannot be shared publicly.

4.7 Conclusion

The data collected and analyzed has provided clues about possible patterns and signals in the behaviour being examined but formal assertions about clear correlations cannot be made as the preceding three analytic methods have shown. Causal linkages can be asserted in error from correlations that have not been carefully evaluated (Myint et al., 2017). Growing volumes of data and access to complex modeling and analytic tools have increased the temptation to stretch explanatory claims for discovered patterns (Benjamin et al., 2018). New and complex information benefits from plural and diverse methods when looking for patterns (McShane et al., 2017). The most extreme individual cases suggests that low movement and low social capital may correlate at the low-income periphery but not when income based clusters are averaged. Outliers can be signals of important patterns but may just be anomalies. Further research focusing on income groups above the 75th percentile and below the 25th percentile would be required to reach a more definitive conclusion. Larger data sets would allow a wider potential range of detectable patterns among social capital and spatial variables. Finally, it is possible that new analytic techniques arising out of statistical and machine learning research will provide greater clarity about information contained in spatial data sets that are formally associated with social features such as those represented by the trust and social networks of social capital (Adams, 2001; Brondizio et al., 2009).

Human activity in urban settings is a continuous and deep synthesis of our relationships with other people in a wide array of places and spaces reflected in everyday life. These places and spaces certainly include virtual and digital environments but even those places cannot escape the tangibility of actual physical locations – servers, computers, devices we hold, the other people we interact with online – all of these things occupy and order the spaces of our lives individually and collectively. The design of physical space at all scales matters greatly to us. Individuals shape their spaces at a personal level – rooms, workspaces – but play a smaller (and less direct) role in shaping the reality of their urban environments as the scale of space increases.

Planners are institutional-scale actors and at that scale, create the contexts within which individuals live and act. If examination of social and spatial remains individual, the importance of the foregoing research would be of limited use to planners in the scope of their institutional responsibilities. At a neighbourhood scale, however, planning and the individual come into direct and meaningful contact. Given the institutional (planners) and individual (citizens) interactions, I conclude this with an explicit movement back to the nature of physical space as an integral human and social reality. Planning enters the city design process not at a building or single plot scale but at a neighbourhood scale and literacy about those implications situate the preceding technical and analytical work back in the realm of planning practice. I will turn now to an articulation of key principles in this process by means of the theory and practice outlined in the work of Christopher Alexander.

Chapter 5: Implications of Spatial Planning and Social Infrastructure[10]

5.1 Introduction

Effective city building requires a conscious integration of the social and physical infrastructures of our communities. The difficulty of formalizing those interactions has been made clear in the preceding research and analysis. I began with a research approach that formalized both social capital and human movement respectively for statistical and mathematical evaluation and comparison. No clear, simple, durable correlation could be detected within the parameters of the research conducted. Clues about variation in social capital outside of the 25th – 75th income percentiles are there but would need additional data collection. In Chapter 2 I demonstrated that survey data can be used to examine social capital components and create synthetic variables suitable for comparison. There may be refinements or adjusted approaches that could sharpen insight on formalized social capital data. In Chapter 3 I measured GPS data points generated by respondents and formally characterized those patterns so that participants could be compared and grouped together to reveal aggregated features. The comparison of both of these modes of data collection and analysis was completed in Chapter 4 using three very different analytic techniques to increase the chance of finding meaningful data patterns. Given the opportunity to return to this area of investigation, it would be useful to capture a larger group of people at above and below these centralized percentiles. The value of detecting social capital and spatial correlations at the more extreme edges, however, reduces the value of the conclusions since the minority of people will fall in those external ranges. Although this is a difficult conclusion to reach, it affirms what the review of social capital and spatial dynamics in Chapter 1 revealed: Social capital is never just social but shapes and is shaped by the myriad cultural, historical and therefore physical realities of environments designed, built and occupied by humans.

Capturing data from participants proved very difficult in terms of recruitment and this was reflected in the raw calling numbers when compared with completions. The custom software and hardware I developed for my own GPS devices proved invaluable in understanding the technical aspects of GPS

[10] This chapter published as: Friesen, M. (2019). Planning for Social Environments: Social Capital in the Context of Critical Realism and the Dynamics of Complex Systems. *Policy and Complex Systems*, 5(2), 221–242. https://doi.org/doi: 10.18278/jpcs.5.2.8

data but the relative cost and high loss rate means that larger scale goals would be costly using that approach. Hardware prices continue to decrease so that per-unit pricing is now less than 50% of what is was in 2016. Over the last few years, low power wide area networks have shown promise for collecting positional data independent of mobile devices or GPS by using ground transmitters and receivers that do not log data but pass it to a local receiver/server (Petäjäjärvi et al., 2017; Petroni et al., 2018; Qadir et al., 2018). This would mean availability of data for participants even if the devices were not returned. The local scale of these networks using devices that are like car key fobs could be valuable for neighbourhood scale location and survey combination approaches at much higher participant number levels (they are also much more energy efficient than GPS devices with weeks or months of duty between charging or battery replacement). It should be noted that since 2016 when the data was collected, more data is available via mobile devices. However, social capital content such as that reflected in the SCGSS survey questions has not been broadly replicated and remains nearly as scarce today as it was five years ago. As identified above, Facebook has made some anonymized data sets available that link transportation choices and nodes with relationship networks, but this data has significant numbers with far less social capital content. While spatial analysis continues to grow, the particular linkage between social capital and spatial patterns at the neighbourhood scale is a problem that needs further research in order to scale across a city or region.

Planning cannot wait until we have fully plumbed these complexities. Instead, provisional insights must be apprehended and put to work while formal understanding is pursued in partial ways through the long labour of experimental discovery. I propose in this chapter how planners may hold in tension our limited but growing knowledge of social and spatial interactions while simultaneously aligning theory and practice to compliment that knowledge. I argue that Christopher Alexander's theoretical and applied architectural and planning work is ideally suited for this formal reflection on space and social relations. In the first half of this chapter, I examine the way Christopher Alexander's work, *The Nature of Order* frames such integration. The range of issues identified in Section 1.6 – negotiated difference, perception of space, digital space, the concept of encounter, neighbourhood tensions, in-group, and out-group dynamics, as well as radicalization and powerlessness – can benefit from the significant synthesis Alexander has developed over the course of his work. In the second half, I extend those insights, returning to the assertions made in Section 1.8 above, that planning must pursue a much deeper understanding of complexity science and networks if the social well-being of citizens is to be responsibly taken up.

One of Alexander's core contributions to a synthetic understanding of the social and spatial is a critique of the reductive simplification and segmentation of many planning and design processes, particularly in their institutional forms. His pivotal essay, *A City is Not a Tree* ((Mehaffy & Alexander, 2016), argues convincingly that living and vital systems overlap richly while highly controlled, standardized systems deliberately reduce overlaps, leading to fragility, including relational impoverishment. Alexander argues that we can avoid this risky simplifying logic by noting seven essential principles for spatial use and human well-being (Alexander, 2003a, p. 3):

1. Form vs. function discussion represents an artificial separation;
2. Planning always has a context that is deep, critical and cannot be ignored without great cost;
3. Planning is effective where design generates unity;
4. Effective planning never marginalizes human feeling because human response is an objective reality that must be addressed;
5. Responsible planning acknowledges that ecology and biology are not fads; they are integral aspects of context, unity, human feeling and other critical variables;
6. Equitable planning fully understands that social agreement is the only way that the complex processes of building and sustaining a city or region can succeed;
7. Beauty must, ultimately, provide a critical accountability for our planning and building.

These potent assertions pull together the social and spatial (including ecological and environmental aspects) without requiring a fully formed technical solution. Alexander proposes translation of these principles into practice by means of 15 transformations (Alexander, 2003a, p. 22)(Alexander, 2003a, p. 22).

5.2 Design Insight Through Social and Spatial Integration

Theory is not as an activity detached from tangible experience. The worlds of ideas and of built things inform each other directly. Building (or the projects of planning) enacts our thinking and, if approached in the right way, refines and extends it. This interaction is at the heart of the primary research question that has guided my study: *What is the relationship between how a statistically significant group of people move around their neighbourhoods, reflected in GPS spatial data, and social capital levels as established by the SCGSS baseline measurements?* Social capital is entirely social while moving around a neighbourhood or city is very much physical – these cannot be divorced from each other though one or the other may be more or less marginalized. Alexander has made a sterling contribution to the nature of rich integration and these insights are reflected in the 15 transformations that embody the tensions of holistic design (Alexander, 2003a, pp. 5–22):

1. *Wholeness and Value as a necessary part of any complex system.*

2. *An intuitive model of wholeness as a recursive structure.*

3. *A mathematical model of wholeness identifying wholeness as a well-defined recursive structure of a new type.*

4. *Objective measures of coherence in complex systems, and the unavoidable relationship between structure, fact, and beauty.*

5. *Fifteen geometric properties [transformations] as necessary and inevitable geometric features of reality in any complex system.*

6. *A meeting point between cognition and objective reality.*

7. *A new, experimental way of determining degree of coherence, degree of life, and relative value.*

8. *The science of complexity must make room for subjectivity, not in the sense of idiosyncrasy of judgment, but as a connection to the human being.*

9. *Local symmetries and sub-symmetries.*

10. *Deep adaptation as a central concept in complex system theory and in architecture.*

11. *The absolute necessity for successful adaptation to be achieved by generative means.*

12. *The effect of structure-preserving transformations on the world and their role in the unfolding of wholeness.*

13. *The hugeness of configuration space and the way the trajectory of a complex system can reach adaptation.*

14. *More on adaptational success as a special kind of trajectory through configuration space.*

15. *Wholeness-preserving transformations are the primary ways the trajectory of a complex system is able to reach successful adaptation.*

Far from being irrelevant, these principles give room for meaning, repeated attempts that improve what we do, mathematical formality in balance with other ways of knowing, the value of human experience both collective and individual, and continuity in its many forms. We may ask: Which of the preceding 15 transformations is purely social or purely physical? They clearly show the inadequacy of such a polarizing question. Holding so many questions in one's mind while working out planning problems or challenges may be expecting too much. I propose that the language and structure of Alexander's ideas become more useful for social and spatial integration if they are translated through an easier-to-use framework.

5.3 Fifteen Transformations into Five Questions for Interrogative Planning

The 15 transformations can be reframed for application by means of five questions that reflect the principles while allowing working planners to integrate social and spatial dynamics more readily (Figure 5-1).

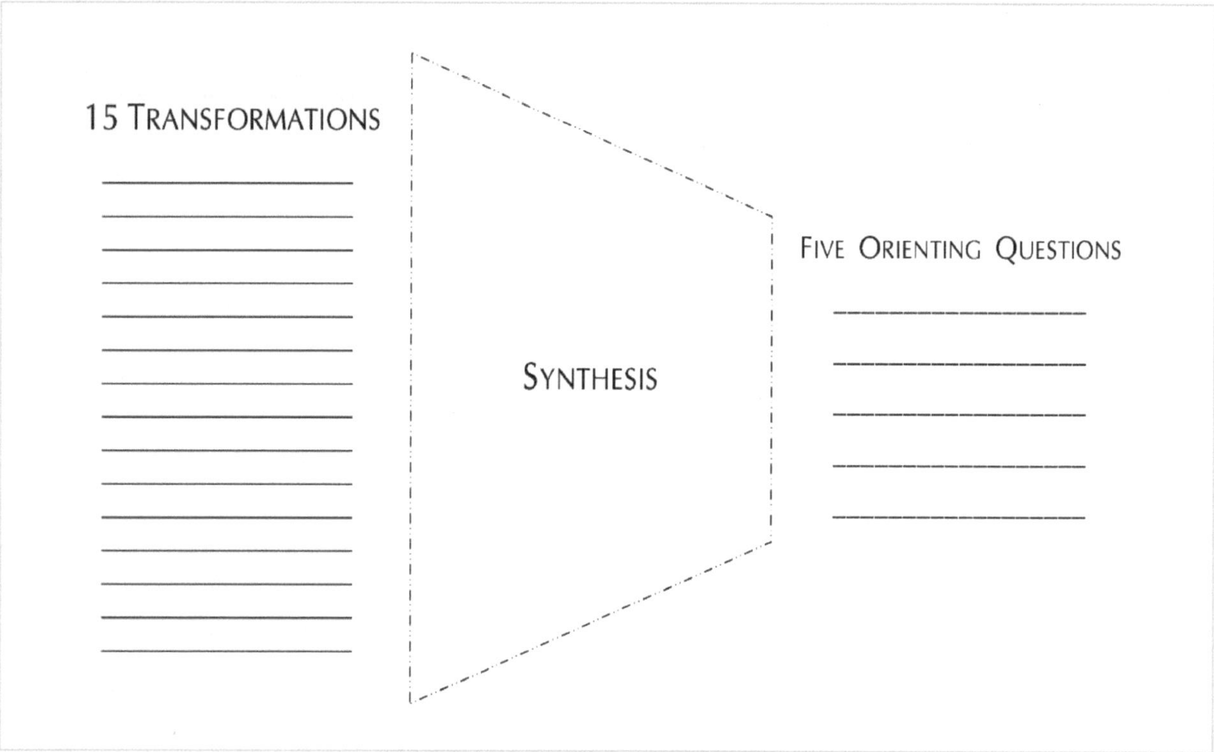

Figure 5-1 - Alexander's 15 transformations can be usefully synthesized into 5 orienting questions. The transformations, synthesis and orienting questions can be used as a framework for model specification and design considerations depending on context.

The value of these structuring questions is that they build valuable tensions into the socio-spatial design process, including both theory and practice. This *interrogative* planning approach invites reflection, engagement, participation, and most importantly continuous revision. They may be used in public settings, committee settings, or any number of other situations where the planning process encounters reflection, conversation, or deliberation. While some strategies and theories favor certain scales (community-based planning tends toward the local and specific, for example), an interrogative approach can move between various scales of functioning structures including the social and spatial. The five questions, which may be asked of both a concrete project as well as a design process are:

- *Is it subjective?*
- *Does it cohere?*
- *Is it adaptive?*
- *Is it recursive?*
- *Is it a functional whole?*

Each of these five questions will orient thinking and practice uniquely depending on what scale is in mind when they are asked. They are, in a sense, scale free, though not in a rigid, over-simplified or reductive way. These questions interrupt simplistic approaches. They allow us to proceed meaningfully and intentionally toward concrete decisions and approaches within the project that is at hand or in mind at a given point in time. I will now turn to an examination of each of the questions and offer a heuristic arrangement to encourage usability. Because they are non-sequential, the five questions can be arranged into the acrostic "*scarf*" where **s**ubjective, **c**ohesive, **a**daptive, **r**ecursive, and **f**unctional whole are represented accordingly.

5.3.1 Is it subjective?

Within the five questions, this is the most easily misunderstood. Planning involves human input across an extraordinary range from the individual citizen to national governments and global trading networks. Alexander differentiates between two types of subjectivity (Alexander, 2003a, p. 13). Type I subjectivity is the one most commonly understood and it is idiosyncratic: I am the only one who holds to a given judgment and my holding that judgment is sufficient to justify it. Type II subjectivity recognizes the individual as a distinct and particular actor who experiences or judges things in a particular way to contribute alongside others. Democratic elections are comprised of many individual decisions that aggregate together to generate a concrete outcome – a particular elected leader. Language use is another example where individuals participate and contribute uniquely in linguistic exchange while being part of a common language (Hintikka, 1998; Quine, 1960). Relational networks are subjective, unique to each person, but when aggregated these relations are part of groups, communities, neighbourhoods and cities – the elements constitutive of social capital at neighbourhood scales. Spatial movement is an open process but, as discovered in Chapter 3, movements of individuals in a city are not at all random – we inhabit and are shaped by complex social/spatial frameworks that occur on pre-existing paths. These paths are denser in proximity to home, our neighbourhoods, and work. The interactions required for Type II community building subjectivity take place in scales small enough for citizen actions and inputs to matter. The pursuit of Type II subjectivity is therefore another form of the good that is essential to a thriving society and is highly relevant for social capital and spatial interactions.

We can only get at this common good, this wholeness, coherence, and related values when "we agree to use the observer's feeling of his or her own wholeness as a measuring instrument" (Alexander, 2003a, p. 13). This tempers individualism while affirming the deep value of human judgment and evaluation – a constant tension for planners. Citizens as citizens lose power as the scale of a social dynamic increases while institutional leaders gain power by means of the infrastructures they direct. Local action, local effects, local voices are a critical layer in building Type II subjectivity – the kind that can enhance human centered value. New approaches to consultation may arise as a diverse range of narrative fragments across a population add up to objective value and a more flexible collective understanding (Mosier & Fischer, 2010; Snowden & Boone, 2007).[11]

Type II subjectivity can also bridge between past and present. It can carry forward what has been learned while simultaneously adapting to emerging technologies, social structures, or ecological dynamics (Arnstein, 1969; Birk, 2016). This bridging is certainly operative in neighbourhood environments and strongly supports one of the secondary research questions (Section 1.4) we have been examining: *How can we improve our ability to measure social capital at a local level (e.g. Census Tract)? What are the challenges or limits to undertaking this scale of measurement?* Where neighbourhood or community engagement is high, Type II subjectivity can operate, reflecting the many aspects coming to bear on a continuously negotiated common good, however tentative that may be. As noted in Section 1.6, digital technology has played a highly influential role in both the social and physical negotiations and represents new frontiers for coordinated subjective development via the rich social

[11] Dave Snowden, for example, has developed the Sensemaker™ software toolkit to capture and meaningfully link a significant range of narrative fragment types into a meaningful evaluation framework – see http://cognitive-edge.com/sensemaker/.

and spatial data flowing through our cities. These digitized collections of Type II subjectivity are worth noting.

Robert Goodspeed (2011) has explored the relative value of different forms of crowdsourcing (he identifies at least six types) for planning settings. New approaches to generative design are emerging in product design, architecture and engineering (Sivram Kish, 2011). Crowdsource planning is a means of approaching concrete needs represented by a distributed subjective group—how the individual's sense of their world can be coordinated with other individuals to arrive at a collectively shaped plan of action (Sennett, 2012). When the relative merits of a few examples such as network science, narrative fragment analysis and crowdsourcing are taken together, it seems likely that we will be able to make use of numerous techniques and methods to develop a more robust subjectivity. As computational modeling becomes more and more ubiquitous, new pathways for collaboration on policy dynamics will be possible outside the realm of specialists (imagine a plurality of models as tools for discussion and decision making by diverse stakeholders).

5.3.2 Does it cohere?

Coherence and wholeness are related and integral to each other. Adaptation considers how much change a functioning system can successfully manage through intelligent recursive feedback. Coherence is about functioning system being dynamically complete in the first place without all of its elements being homogenous. Although there may be many complex and diverse functions within a whole, the interactions and relations are tuned so that they complement each other. Without coherence, a system possesses either too much redundancy or too little interaction. A living system such as a bacterium possesses coherence—proportion, distinction, complementarity, flow of resources, and so on while remaining dependent on other systems outside itself. Alexander argues that coherence plays an indispensable role in designing and building robustly. He argues that new, experimental means of determining degrees of coherence, of life, and of meaningful interaction across scales are critical assets for designers and must be more rigorously explored (Alexander, 2003a, p. 8).

These concepts have historical precedence in human inquiry. Coherence is at work in the guise of elegance in many scientific theories and practices, often functioning as a guide to what is important amid complex and potentially confusing options. Alexander argues that resistance to the idea of coherence as a quantifiable category generally stumbles on confusion over problems of degree and not kind:

> *But there is a very thin line—in fact, I would argue there is no substantial line at all—between the issue of relative coherence of subsystems in a physical-mechanical system, and the more complex distinctions of coherence in an aesthetic entity—the phrasing of a piece of music for example (Alexander, 2003a, p. 8).*

Science has a long history of facts as far as they relate to parts but when it comes to the possibility of aesthetic facts, notions of elegance, seamless integration of form and function, we may have important clues to help us. In mathematics and physics the very large phase spaces they are concerned with may be narrowed by choosing a simple but sufficient explanation rather than a highly convoluted explanation. If a choice between the two is required, the simpler one is generally more useful. Alexander suggests that we should not, in our science, hold to facts about parts alone:

> *After all what is science? It is the study of what really happens, how the world works. Done in such a way that agreement can be forged by clear thought, and by empirical procedures... There is clear thought about structure; and there is empirical basis and procedure specified, which allow people to form agreed on, shared observations, and thereby to reach—at least tentatively and roughly at first—shared understanding and reliable results (Alexander, 2003a, p. 9).*

We may progress through gaining something that is "about right" such that refinement over time is possible. Something that is alive and adaptive, able to respond well to its environment, must be coherent—it must "hang together" in highly sophisticated ways. Planning has at various points approached ideas of coherence through an emphasis on systems-related theory and practice (Allmendinger, 2009, pp. 49–62) although the systems in mind dating from the 1970s were too often focused on simple, manageable systems (information and computing would remove unknowns) rather than complex systems (variables and persistent contingency ensures permanent unknowns, regardless of computational power) which have developed more recently (Allmendinger, 2009, p. 62). Effective design, under those regimes, was equated with reducing the complex to the manageably complicated. Thriving social and spatial systems function as nested systems of relations, networks of energy, food, communication, spatial occupation, movement, and even of ideas. These dynamics hold together, they cohere. In Section 1.6 I argued that the concept of difference, while prevalent in planning theory, has a wide range of meanings. Coherence in a social context means that a common good may have been identified, a balance of interests negotiated. These negotiated differences can include things like allowable uses for parks, which schools or libraries to invest in, and how cultural interests will be prioritized. The principle of coherence supports the creativity and compromise required daily of planners and is more realistic and dynamic than settling for the lowest common denominator where populism has been found to be most active (Rooduijn, 2014). Stability through complex change provides social and cultural continuity so essential for human thriving. Coherence guides such flexibility and leads us to another question concerning adaptivity.

5.3.3 Is it adaptive?

Adaptation in our current study is a means of finding a useful path through the impossibly large configuration space of human societies and physical dynamics in urban settings. Permanent variability seems, at face value, the very thing planners are seeking to overcome. I have tried to understand how spatial movement may correlate with social capital to work out how to map our social structures and gain critical insight about our adaptability during periods of disruption. Our inherited organizational and institutional structures have tended to assume deep social insight would keep pace with technological and bureaucratic efficiencies. We are finding, however, that this is a mistaken assumption which continues to plague us (Alexander, 2003b; Taleb, 2010; Westley et al., 2006; Wheatley, 2006). Research on the social consequences of technological evolution is not linear – our social structures and institutions don't adapt as quickly as the technology does (Arthur, 2009). There is a lag between the metabolic rates of technological change (faster) and institutional development (slower) across all fields and social strata historically (D. W. Allen, 2011). One example is the emergence of Creative Commons licensing in the face of old and inadequate copyright laws, a more adaptive response to a rapidly changing creative space. Planning faces such challenges through encoded laws and regulations that may often lag changing social and spatial needs. This is perhaps most clear where informal developments in cities around the world simply happen without any planning involvement or approval to which planning responds, if at all, after the fact.

Alexander outlines how, in the face of such disruptive evolution, we can devise approaches to design, build and plan that are better able to adjust to change over time. His key insight is that successful change requires frequent small steps that response to changing conditions. Adaptive planning processes can only come about through generative strategies. Generative strategies require adapting to changing conditions. Disruptions may arise from the natural order, social change, or technological innovations. If we are unable to adjust well to unknown future variations, our chances of surviving diminish and along with it our hopes of thriving:

> *Deep adaptation is the process whereby the landscape, or a system, or a plant, or a town, proceeds by a series of spatially organized adaptations in which each part is gradually fitted to the parts near it: and is simultaneously fitted by the whole, to its position and performance in the whole (Alexander, 2003a, p. 15*).

Truly adaptive strategies are generative because they tune to actual conditions including social and spatial conditions. Racial tensions in communities, migration pressures in urban settings, in-group and out-group tensions, as well the presence of radical political groups were noted in Section 1.6 as central themes for the social and spatial deliberations planners are responsible for. These differences, these encounters, are lived out directly every day in cities quite outside the consideration of planners. Where planners do engage in these community dynamics, Alexander argues that "spatially organized adaptations" would involve things like policies, acts, consultations, charrettes, research, public debates, and a host of other dynamic inputs in ways that actually change the planning process. Specialization is valuable but has tended to sever links between micro-scale tuning and macro-scale adaptation, something that occurred in the 1980s approaches to cod stock management on the East Coast of Canada where scientific expertise was accepted though it was counter to local knowledge (Cayley, 2009, pp. 238–256). From a planning theory vantage point, Lindblom and the incremental pragmatists have been, in a general sense, more tuned to the smaller cycles (adjust as you go) while the rational comprehensive method was more tuned to the larger meta-cycles—follow the formalized plan regardless (Lindblom, 1959). We know from our earlier sections that where local agency is low (for both individuals and community groups), social trust and constructive interactions with people who are different tend to be weak. Mapping social capital at neighbourhood scales would provide a resolution closely tied to the possibility of acting, of having agency related to matters such as employment, housing, education, and a host of localized quality of life issues.

Interrogative planning asks us to reflect on both the micro- and meta-dynamics with which a given plan, process, or design problem, must contend. We will need new analytic and process strategies and tools to increase awareness of how adaptive approaches can be nudged toward positive rather than destructive outcomes. Computational approaches such as interactive genetic algorithms (IGAs) suggest that in modeling environments, adaptive processes provide exploratory supports for effective solutions (Goldberg, 1989). Adjusting to unknown dynamics will determine our ability to implement and sustain robust and desirable actions (Benkler, 2006; C. Christensen, 2011; C. M. Christensen et al., 2006).

5.3.4 Is it recursive?

Being recursive requires an ability to understand how information, resources, feedback, novelty, and other flows that drive activities like planning become new inputs back into the process over time. A truly recursive process must be much more than weak feedback for public relations purposes or to fulfill legal requirements (Arnstein, 1969). A recursive structure is an order of magnitude more difficult and nuanced than consultations that reflect one more item on a formal to-do list. A hallmark of a genuinely recursive cycle is feedback designed to change structural qualities of the system itself.

We are accustomed to communication-related feedback through our talking and listening. We plan, act, get feedback in various forms, adjust the plan, keep acting, and so on (Allmendinger, 2009, pp. 197–223; Serres, 1995). A more difficult problem involves thinking about what substantive structural feedback means for built environments, transportation structures, financing, life-cycling, and other practical matters. Alexander explains how the interactions of different orders of organization and function could be designed in just this way:

> *As we know from recursive function theory, surprisingly simple ideas, when applied recursively at a variety of nested levels, can have profound and effective consequences—and, often, surprising ones. Both Ian Stewart and Brian Goodwin acknowledge the important of recursive ideas, and say that in their view, too, recursiveness of qualities is likely to be a feature of all living structures (Alexander, 2003a, p. 7).*

An example from a non-human living system may help to illustrate this. Termites have adapted over a very long time to keep their mounds cool in the unforgiving heat of an African sun. Their secret is that with relatively simple chemical signals iterated over time, a unique physical form results (Jahanmiri, 2014). The host of small adaptations over time are a key to their success. It is not uncommon to see designers, architects, engineers, and planners looking to nature for clues about better design. The problem is that we may often imitate an outcome without understanding the generative process that led to a particular result. New urbanism has often been criticized for creating what looks like a village without all the decades and centuries of human interaction that stand behind an actual functioning village. A model of a leaf, however realistic looking, cannot compare in complexity to a living leaf (Mehaffy & Alexander, 2016). Something that is alive is many orders of magnitude more organized, complex, regulated, and detailed than any mechanistic (or bureaucratic) system we can devise either in practice or theory. Alexander argues that we need to move recursive insights deeper into the process – including the social and spatial interaction that concern us here. It is more difficult to design dynamic feedbacks between and among scales, but this continuum of scales and iterations is essential for planning.

The nettle of complexity to be grasped can only be managed if the system itself can tune and adjust both in detail and in substance. If we over-manage this complexity, paralysis, or unintended consequences result. This has been the classic shortcoming of various control-driven structures and approaches—they over-constrain as they try to control. More effective recursive processes could arise from social needs and dynamics (such as changing social capital levels) to influence physical planning decisions more strongly such as housing types, modal developments in transportation, and zoning suited to the uses and needs required by a community. For example, an individual who believes that his or her community perceives them negatively is less likely to contribute to a common good. As noted in Section 1.6, withdrawal, marginalization, and a propensity to adopt more radical ideas is a possible consequence of feeling distanced from meaningful input. Finding ways for recursive processes to build social and physical capital is essential for effective planning. Social capital monitoring can provide the clues needed to build accurate pictures of the social state of our changing communities across a city.

Organizational systems may also reflect certain recursive functions that limit their growth or development. Strong, nearly closed systems like bureaucracies may eliminate or suppress anything that doesn't fit within the parameters of the system even when those changes would lead to much better outcomes (Christensen, 2011). Investments in understanding and implementing recursive thinking can help planning to avoid ineffective development (Forrester, 2009; Trofimova, 2000).

5.3.5 Is it a functional whole?

The fifth question involves an assessment of wholeness: does the designed/planned/built entity we have in mind function, does it work? Wholeness represents the fullest integration of all the dynamics that impact human society. A functional whole is an identifiable organism, entity or spatial arrangement that integrates a wide range of diverse dynamics to produce novel effects, direction or intention that is valuable. Being a functional whole is deeply integrated with the other four questions. If there is no preservation of a functional whole amid change, the system or entity cannot evolve, and it will simplify relative to its environment and break down. When systems collapse, they may do so partially and re-organize more effectively in a new form better suited to the new conditions (catagenesis) or they may collapse to a point where they cannot recover their former levels of complexity(Homer-Dixon, 2007). From an urban planning vantage point, the challenge of maintaining a functional whole could mean the preservation of an existing neighborhood through the process of renewal, development, or disruption. In some cases, less effective solutions simply over-write pre-existing social and physical infrastructure, often at significant human, cultural and ecological cost (Cheung & Leung, 2012). Actual conditions in a neighbourhood are one of the most telling measures of planning efficacy (or failure) and social capital has long played a central role in that evaluation (Hanifan, 1916). Continuously occupied cities that have thrived over centuries provide clues about how this principle works.

For healthy growth to happen, a functional whole must be maintained during the changes. If, as noted above under adaptation, the changes are too great and functioning stops, growth will also stop – or it will become destructive growth, unsustainable and harmful. Maintaining a functional whole or bringing a functioning whole into existence is a deep challenge for designers and planners. Alexander argues that it is perhaps the most significant challenge we face:

> *How can a complex system find its way to the good configurations? In a theoretical sense, we may say that the system walks through configuration space, taking this turn and that, and always arriving at a well-adapted configuration. The huge question, of course, is: How this walk is controlled: what are the rules of the walk, that make it lead to good adaptation? Although a few very preliminary answers have been given to this question, no good ones have yet been given. This is perhaps THE scientific question of our present era (Alexander, 2003a, p. 19).*

A key aspect of a functioning whole is that it interacts in a good way with both the system that it lives inside as well as with the smaller systems that give rise to it. This notion of cycles of function and wholeness within and supporting other kinds of wholeness was noted by ecologist C. S. Holling who observed how cycles of growth and decay at all scales had regular patterns that yielded endless novelty including both human and natural systems (Gunderson & Holling, 2002). Alexander argues that we have not learned to assess wholeness or how to preserve it amid the changes we make in planning and building:

> *Nature has a way—built into the majority of systems—of finding its way to a well-adapted state for any given complex system (at least for most cases). We do not have a way. For buildings—and indeed for any complex system [e.g. Cities]—this is the most fundamental practical issue of all* (Alexander, 2003a, pp. 19–20).

Our interventions often simplify existing dynamics through a desire to control and manage them. The human factor in cities is critical and wholeness attends to serving the needs of people without losing sight of the other scales of interaction needed for that to happen—climate, ecology, energy security, food supply, transportation, cultural vitality, freedom of speech, justice, and so on, all matter. Improving our ability to see and understand social and spatial pattern interactions would greatly increase our capacity to iteratively design for greater community health.

Interrogative planning and Alexander's approaches to developing and building have powerful social/cultural dimensions that are inseparable from contemplations about the built environment. Linguists, technology scholars, organizational theorists and many others have noted that what we become is both reflected in what we make and in how what we have made then influences us (Postman, 1993).

5.4 Disciplines for Integrating Social and Spatial Research and Practice

I will now turn to three core disciplinary themes that can integrate social and physical practice for planners: Deeper capabilities in network theory and mathematics; More adaptive and effective physical infrastructure; Improved data sources and data structures. The specific gains we make in understanding social and spatial dynamics for human flourishing will need to be supported by greater investments in these disciplinary themes.

5.4.1 Network Theory and Mathematical Capability

Another vital bridge between the practice and theory of social and spatial interactions is network science. In Chapter 1, the importance of networks as a bridge between complexity science and social capital was made clear – social research requires network science as does the study of human movement in urban spaces. Network science permeates all five of the interrogative planning questions: Subjectivity relates individual or sub-scale inputs to objective outcomes, coherence requires finding meaningful interactions between disparate elements, adaptive and recursive processes are comprised of relationships between social, physical and perceptual entities, and functional wholes can be seen as networks where a range of tensions and trade-offs lead to a valuable and useful entity, whether that is a healthy neighbourhood or a thriving city. Network science literacy is, therefore, indispensable for both formal researchers and professional practitioners in a wide range of contexts including planners.

Mathematical advances in network science have allowed for sophisticated relational mapping across many fields – something that is indeed of value for planners since planning has a great deal to do with how spaces, places, people and resources relate to each other. Mathematics provides the formalized environment for network exploration. Network topology – the structural properties of a network – is reflected in the social capital questions about the nature and number of family, close friends and acquaintances. Graph theory (early social networks were called *sociographs* or social graphs) has roots in Euclidean geometry extended by mathematicians like Leonhard Euler (Debnath, 2010). Computer science has both benefited from and advanced this study. Many of the analytic tools used in this study rely on these formalized relationships applied through computer algorithms. Disciplines such as engineering, biology, physics, operations research, logistics analysis, human geography, and environmental research reflect extensive use of network analysis. These disciplines are part of the planning environment and intersect significantly with the highly diverse range of disciplines that the review of social capital reflected in Chapter 1.

Network data in urban settings is being generated at remarkable rates – GIS information, traffic flows, pedestrian patterns, commercial transactions, resource use (and disposal) measures, conservation mapping, tagging of images, online comments and interactions, embedded devices both mobile and static, and a host of other flows operate intensively in cities. Data is related to other data in networks that can be studied from many angles. One of the hoped-for results is that investments such as this study will lead to the design and development of enriched network analysis in support of more resilient cities (Boyd & Folke, 2011; da Silva & Morera, 2014; Hatt, 2013; Serre et al., 2012). Adaptive cities that are responsive to change require administrative insight that is tuned to the dynamic understanding and analysis of elements such as the social interactions that are present everywhere (Benkler, 2006; Castells, 2009; Pentland, 2014; Wellman & Rainie, 2012).

One implication of network dynamics and topology is that we cannot treat elements in discrete ways – nodes are connected by a dynamic of some kind and knowing lots of information about a node (e.g. Census data about individuals) is only a very partial kind of knowing. We know far more specific things about individuals (age, income, preferences) than we do about how those individuals relate to those around them and the patterns that those interactions form.[12] Data collected integrated personal data with network-based social and spatial data. The relational network revealed in friends, family and acquaintances was self-reported. Organic data about social networks through social media is pervasive and very useful where available but the strength of this study was an experimental method that looked for correlations between those social and spatial patterns. Finding formal correlations through careful methodological experiments will increase the value of the large quantities of socio-spatial data collected by organizations like Facebook (Bailey et al., 2018). Based on these formal uses of network architecture, planners can inform intelligent policies and practices including the significant challenge of social impact measurement (Aaltonen, 2007; Antadze & Westley, 2012; Feldstein & Sander, n.d.; Friesen, 2013a). Enriched understanding of all the networked implications of these changes – from the social to the spatial – is a long-term growth need for planners.

5.4.2 Adaptive and Effective Physical Infrastructure

Since the 1960s municipalities have used computers (Alexander & Manheim, 1962) to manage their GIS data in increasingly sophisticated ways. Adaptive cities will need to consider not only their responsiveness to economic, environmental, population, housing and communications changes from a

---

[12] In *A City is Not a Tree*, one of the most significant critiques that Alexander makes of contemporary planning management and bureaucracy is how they instinctively reduce relational overlaps to increase control and reduce variability. The effect of this is expressed in a formal mathematical comparison. If 20 people (n=20) are formed into as many groups as possible, a simple rule change radically affects the potential range of options. If no overlaps are allowed, i.e., you cannot be a member of two different groups, the maximum number of groups is 20-1 = 19. If, however, multiple groups per person are allowed, the maximum number of groups is 2^{20} = 1,048,576. Living organisms are comprised of highly overlapped systems and interactions which reflect the 2^n rather than an n-1 arrangement. This significant richness contributes to the challenges we face computationally when network structure and dynamics are analyzed leading to many non-computable problems.

human vantage point, they will also need to do better at understanding how resilient and adaptable their physical infrastructure is. As physical infrastructures interact with social infrastructures shaping human behaviour, new phenomena – some good, some less desirable – emerge that act back on the original design. Studying the social ability of physical spaces is not new (Nj Thrift, 1983) but formal planning has often been driven by structural land, transportation and legal frameworks ahead of sociability. We appear to be reaching a point in our communities where we can ill afford to ignore the long-term costs of a weakened social fabric (Bethune, 2014, 2014; Dunkelman, 2014; Marche, 2012).

Alexander's idea of subjectivity may seem counterintuitive but simple observation confirms that it is possible (and even common) to build something collectively for which no master plan exists and which, even after it is built, is not known in its entirety by any one person. The Internet is a vast collection of devices, wires, radios, and computing nodes so extensive that the exact configuration is unknown except by estimate. The Internet is, in this sense, a human version of a termite mound where no single directing entity organizes the physical space. To learn about what we have built, we must conduct experiments on the network that measure nodes and edges (Newman, 2010, p. 3). It is easy to conceive of many aspects of cities as rich representations of this partially ad hoc phenomena. Some elements of the city are better known than others and administrations go to great lengths to build databases describing where water, sewer, utility, and other services are located. This data is, however, never fully complete and old infrastructure may only be generally known, encountered when an unmarked water line fails. If we extend our field to include dynamics on these networks and go yet further to include human factors, it becomes very sensible to think that even in the most advanced cities, we only have, at the best of times, partial knowledge of the state of a city at any given time. Informal developments are an even more clear example of this dynamic of life in cities that exist without master plans (Goldstein et al., 2009). Some of our most significant gains in planning will come from studying the social dynamics of our cities.

5.4.3 Improved Data Sources and Structures

Traditional sources of data include census, market research, a host of municipal records, maps, GIS, and community records. Commercial data about spending patterns, customer profiles, and market characteristics are rich but often not available to researchers. Mobile device and transaction data are significant continuous sources of information flow that would be of use to planners and amenable to a wide variety of useful network analyses but are usually off limits as well. Planners will need to find ways to remain current about data sources, both traditional and emerging, and to access as much of that data as possible to improve their ability to sense and respond to changes that are less obvious than physical, land-use or policy changes (Roo & Silva, 2010). New, carefully designed research will lead to new data sources that will help to answer questions about complex socio-spatial dynamics such as changes in social capital at neighbourhood resolutions. New flows of data feeding dashboards, analytical tools, decision-support frameworks and policy directions show no sign of slowing down. I am hopeful my work will contribute to the movement from data to information, from knowledge and understanding.

The relational dynamics of individuals and groups was an early application of network data and exploration although acquiring the data was not easy (Smith & Moreno, 1951). Acquiring good social data can be difficult but social insight remains critical for planners when undertaking analysis of potential projects and evaluating the impact of completed projects. For example, relational data will be an essential part of answering questions such as which investments of limited social service funds yield the greatest common good contribution in a given neighbourhood? It remains far easier to measure inputs in a system than it does to quantify in a reliable way what the social impact of a particular investment is (L. Nguyen et al., 2015) and this will continue to be difficult without deep data capabilities.

Though not typically within normal planning considerations (except for disaster preparedness), the social realities of epidemics is a matter of significant concern for anyone who is involved with managing or designing cities. Social networks have a significant impact on the ways in which diseases spread, grow, and decline in given populations. The SARS outbreak had an immediate and direct impact on Toronto and other cities that had to contend with changing patterns in tourism, city mobility, medical capacity, and public relations matters (Government of Canada, 2005). Fourteen years later, an enormous exclamation mark was added as COVID-19 swept not only Toronto but the world. It is virtually impossible to understand, monitor or predict outbreaks among social populations without network data. Although planners may not typically encounter outbreak scenarios, COVID is providing an example of how knowing the social landscape of a city is critical for effective policy response. Epidemiologists and populations researchers are important sources of intelligence for municipal planners who may want to understand relational dynamics for other reasons – such as the spread of ideas and influences. The mobilizing of ideas, sentiments, and information follows pathways similar to contagion networks and these contagion networks are used to understand more about the social fabric of a community. This cross-disciplinary linkage may be under-utilized by planners and could be remedied through network science study and application in core training and development settings.

As our global interconnections intensify, one of the measures that will gain increasing attention is the matter of resilience – how well does a given network or system respond to disturbance wither internal or external. Gross capacity of a system or network (say Gross Domestic Product) may end up being a secondary indicator while resilience is the more critical measure because it signals how capable of absorbing disruption a given system is (da Silva & Morera, 2014).

With investments in research, exploration and training, planners will become more adept at seeing the social landscape of the city or community and based on those patterns, adjust development, infrastructure, social and other program recommendations in such a way that the common good is more fully served and costly mistakes caused by unforeseen consequences. Enriched public engagement, process changing feedback loops, and more responsive planning approaches will all require improved network approaches.

5.5 Conclusion: The Future of Planning as a Synthesis of the Social and the Spatial

Interrogative planning is a *higher* order of planning, a *more* rigorous administration and a more mature form of action than overly simplified or over-controlled forms of design. One of the research questions raised in Section 1.4 asked: *What are the implications of improved, behaviour-based measurements of social capital levels for urban and social planners?* From a variety of angles, I have articulated that the demand for increased social and human vitality is integrally linked with the interactions of social and spatial conditions. I have also argued that trying to understand these dynamics will not even begin to occur if we do not take up non-linear, complex dynamics as central. This means engaging directly with fundamental unknowns and resisting many built in habits of planning practice that seek high levels of control or formalization. Over-controlled design can mimic the appearance of a good city as noted above in the example of a wax leaf vs. a living leaf. Designing cities and communities that maximize living vitality requires growth in the directions suggested by the five questions presented here and the deeper logic of Alexander's 15 transformations. If pursued the rich interactions of social and physical space in cities will be much more adequately met even where research results are only very partial. Perhaps one of the unintended benefits of our failure to fully control urban dynamics is that the city has room to evolve in its own way:

> *A city is, perhaps, closer to being a self-organizing system than a more conscious artificial creation such as an entire nation, which is the product of a particular historical moment and which may yet prove to be of only transient importance...Over time [a city] must continually adapt and find new ways to survive, and it must be able to switch from one technology to another to do it* (Burdett & Sudjic, 2011, p. 48)(Burdett & Deyan, 2008, p. 48).

Public service increasingly requires that recommendations and accountability are rigorous, empirical, and solidly underwritten by what is quantifiable. Plans need to have maximum positive impact with a minimum financial, social and environmental cost including a positive response to disruptive variables. Urban intensification requires increased investment in the social implications of spatial plans. In the face of seemingly intractable complications, the solution is not to eliminate rigour but rather to enrich it with a fuller understanding of what is taking place in a given setting – will a new subdivision negatively impact social capital levels? Will it weaken the cultural, social, or institutional life of a given area? What are the costs of a given project including social costs? How might that be calculated or estimated? Separating land use issues from social issues has often led to unintended negative consequences and failures (e.g. 1950s housing reform, downtown freeway development, removal of public transit infrastructure in the 1960s)(Sugrue, 2005). Our complex urban settings are increasingly interdependent and it is difficult to imagine this trend changing (Bak, 1996; Homer-Dixon, 2007). Contending successfully with these dynamics at large scales, or where data flows are significant, will require some degree of mathematical insight deployed in conjunction with network and computational sciences. Alexander and others fully embrace this aspect of rigorous design.

It is certainly worth considering that a very different kind of approach, one that is more mature, robust, rigorous and effective, is at the very least possible and may meet new urgency in an increasingly urban-intensive and world. In course of a single year we have witnessed the attempts to control a disease which did not follow simple paths socially or spatially. New flows of data, social network tools, network science, computational modeling, big data and emerging platforms may radically alter the social structures with which urban planning contends. The convergence of rapid technological change in both the built environment and the social/institutional environment will require planning approaches that are better suited to these dynamic environments both in times of normal function as well as during seasons of widespread disruption. Again, if the social and spatial are recognized as deeply related, and if the principles proposed by designers who understand complexity are adopted, quality of life in normal times and resilience in the face of disruption will both be met. We are constantly tempted to simplify in order to control. Studying the complexity of the social and spatial while making provisional gains through Interrogative planning can bridge and support urban planning at a time when the demand for rich integration to solve our most pressing challenges rises daily.

References

Aaltonen, M. (2007). *The third lens: Multi-ontology sense-making and strategic decision-making*. Ashgate Publishing, Ltd.

Aaltonon, M., Barth, T., Casti, J. L., Mitleton-Kelly, E., & Sanders, T. I. (2005). *Complexity as a Sensemaking Framework for Methodology* (FFRC-Publications 4/2005). Finland Future Research Centre.

Abbott, R. (2009). The Reductionist Blindspot. *Complexity, 14*(5), 10–22. https://doi.org/10.1002/cplx.20274

Adams, R. M. (2001). Complexity in Archaic States. *Journal of Anthropological Archaeology, 20*(3), 345–360. https://doi.org/10.1006/jaar.2000.0377

Adler, P. S., & Kwon, S.-W. (2002). Social Capital: Prospects for a New Concept. *The Academy of Management Review, 27*(1), 17–40. https://doi.org/10.2307/4134367

Adrienko, N., & Adrienko, G. (2011a). Spatial Generalization and Aggregation of Massive Movement Data. *IEEE Transactions on Visualization and Computer Graphics, 17*(2), 205–219. https://doi.org/10.1109/TVCG.2010.44

Adrienko, N., & Adrienko, G. (2011b). Spatial Generalization and Aggregation of Massive Movement Data. *IEEE Transactions on Visualization and Computer Graphics, 17*(2), 205–219. https://doi.org/10.1109/TVCG.2010.44

Alexander, C. (2003a). *New Concepts in Complexity Theory Arising from Studies in the Field of Architecture: An overview of four books of The Nature of Order with emphasis on the scientific problems which are raised*. Katarxis. http://www.katarxis3.com/SCIENTIFIC%20INTRODUCTION.pdf

Alexander, C. (2003b). *The Timeless Way of Building*. Oxford University Press.

Alexander, C., & Manheim, M. (1962). *The Use of Diagrams in Highway Route Location: An Experiment.* (Research Report 161 / R62-3). Department of Civil Engineering.

Allen, D. W. (2011). *The Institutional Revolution: Measurement and the Economic Emergence of the Modern World*. University of Chicago Press.

Allen, T. J., & Henn, G. (2006). *The Organization and Architecture of Innovation: Managing the Flow of Technology* (1st ed.). A Butterworth-Heinemann Title.

Allmendinger, P. (2009). *Planning Theory*. Palgrave Macmillan.

American Planning Association. (2012, December 6). What is Planning? [Official Website]. *What Is Planning?* https://www.planning.org/aboutplanning/whatisplanning.htm#5

Amin, A. (2002). Ethnicity and the Multicultural City: Living with Diversity. *Environment and Planning A: Economy and Space, 34*(6), 959–980. https://doi.org/10.1068/a3537

Anderson, P. W. (1972). More Is Different. *Science, 177*(4047), 393–396. JSTOR.

Antadze, N., & Westley, F. R. (2012). Impact Metrics for Social Innovation: Barriers or Bridges to Radical Change? *Journal of Social Entrepreneurship, 3*(2), 133–150. https://doi.org/10.1080/19420676.2012.726005

Anthopoulus, L. G. (2015). Understanding the smart city domain. In *Understanding the smart city domain: A literature review* (pp. 9–21). Springer.

Archuleta, A. J., & Miller, C. R. (2011). Validity Evidence for the Translated Version of the Personal Social Capital Scale Among People of Mexican Descent. *Journal of the Society for Social Work and Research, 2*(2), 39–53. https://doi.org/10.5243/jsswr.2011.2

Aristotle. (2012). *Aristotle's Nicomachean Ethics* (R. C. Bartlett & S. D. Collins, Trans.). University Of Chicago Press.

Arnstein, S. R. (1969). A Ladder Of Citizen Participation. *Journal of the American Institute of Planners, 35*(4), 216–224. https://doi.org/10.1080/01944366908977225

Arthur, W. B. (2009). *The Nature of Technology: What It Is and How It Evolves*. Free Press.

Axhausen, K. W. (2008). Social networks, mobility biographies, and travel: Survey challenges. *Environment and Planning B-Planning & Design, 35*(6), 981–996. https://doi.org/10.1068/b3316t

Bailey, M., Cao, R., Kuchler, T., Stroebel, J., & Wong, A. (2018). Social Connectedness: Measurement,

Determinants, and Effects. *Journal of Economic Perspectives*, *32*(3), 259–280. https://doi.org/10.1257/jep.32.3.259

Bailey, M., Farrell, P., Kuchler, T., & Stroebel, J. (2019). *Social Connectedness in Urban Areas* (Working Paper No. 26029). National Bureau of Economic Research. https://doi.org/10.3386/w26029

Bak, P. (1996). *How Nature Works: The Science of Self-Organized Criticality*. Springer-Verlag.

Baker, W. E. (1990). Market Networks and Corporate Behavior. *American Journal of Sociology*, *96*(3), 589–625.

Bechard, M., & Marchand, I. (2006). *General Social Survey Cycle 19: Time Use(2005)* (Catalogue no. 12M0019GPE). Statistics Canada. http://data.library.utoronto.ca/datapub/codebooks/cstdli/gss/gss19/gssc19gidv2.pdf

Beer, D., & Burrows, R. (2013). Popular Culture, Digital Archives and the New Social Life of Data. *Theory, Culture & Society*, *30*(4), 47–71. https://doi.org/10.1177/0263276413476542

Bell, G., Boellstorff, T., Gregg, M., Maurer, B., & Seaver, N. (2015). *Data: Now Bigger and Better!* Prickly Paradigm Press.

Belliveau, M. A., O'Reilly, C. A., & Wade, J. B. (1996). Social Capital at the Top: Effects of Social Similarity and Status on CEO Compensation. *The Academy of Management Journal*, *39*(6), 1568–1593. https://doi.org/10.2307/257069

Benjamin, D. J., Berger, J. O., Johannesson, M., Nosek, B. A., Wagenmakers, E.-J., Berk, R., Bollen, K. A., Brembs, B., Brown, L., Camerer, C., Cesarini, D., Chambers, C. D., Clyde, M., Cook, T. D., De Boeck, P., Dienes, Z., Dreber, A., Easwaran, K., Efferson, C., … Johnson, V. E. (2018). Redefine statistical significance. *Nature Human Behaviour*, *2*(1), 6–10. https://doi.org/10.1038/s41562-017-0189-z

Benkler, Y. (2006). *The wealth of networks: How social production transforms markets and freedom*. Yale University Press.

Bennett, L., & Vale, F. (2015). *Spatial Statistics: Simple Ways to do More with Your Data* [Esri Technical Workshop]. http://www.esriurl.com/spatialstats

Berger, M., Wong, L., & Rhode, I. S. of D. (2014). *Resilience and Adaptability*. Birkhäuser. https://www.degruyter.com/view/product/428107

Berger-Schmitt, R. (2002). Considering Social Cohesion in Quality of Life Assessments: Concept and Measurement. *Social Indicators Research*, *58*(1/3), 403–428. https://doi.org/10.2307/27527016

Berkman, L. F. (1986). Social Networks, Support and Health: Taking the Next Step. *American Journal of Epidemiology*, *123*(4), 559–562.

Berkman, L. F., Melchior, M., Chastang, J.-F., Niedhammer, I., Leclerc, A., & Goldberg, M. (2004). Social Integration and Mortality: A Prospective Study of French Employees of Electricity of France–Gas of France: The GAZEL Cohort. *American Journal of Epidemiology*, *159*(2), 167–174. https://doi.org/10.1093/aje/kwh020

Bernard, H. R., Killworth, P., Kronenfeld, D., & Sailer, L. (1984). The Problem of Informant Accuracy: The Validity of Retrospective Data. *Annual Review of Anthropology*, *13*, 495–517.

Bethune, B. (2014, August 8). The End of Neighbours. *Macleans.Ca*. http://www.macleans.ca/society/the-end-of-neighbours/

Bibby, Kent, Kanti, & Mardia. (1981). *Multivariate Analysis* (1st ed.). Academic Press.

Birk, R. H. (2016). Infrastructuring the social: Local community work, urban policy and marginalized residential areas in Denmark. *Environment and Planning A*, 767–783. https://doi.org/10.1177/0308518X16683187

Bivand, R. (2013, January 22). *CRAN Task View: Analysis of Spatial Data* [R Packages]. CRAN R-Project. http://cran.r-project.org/web/views/Spatial.html

Blokland-Potters, T., & Savage, M. (2008). *Networked urbanism: Social capital in the city*. Ashgate Publishing, Ltd.

Blondel, V. D., Guillaume, J.-L., Lambiotte, R., & Lefebvre, E. (2008). Fast unfolding of communities in large networks. *Journal of Statistical Mechanics: Theory and Experiment*, *2008*(10), P10008. https://doi.org/10.1088/1742-5468/2008/10/P10008

Boase, J., & Wellman, B. (2004). *Suggested Questions on Social Networks and Social Capital* [Report to the Policy Research Initiative]. Policy Research Institute (now Policy Horizons Canada).

Boin, A., Ekengren, M., & Rhinard, M. (2020). Hiding in Plain Sight: Conceptualizing the Creeping Crisis. *Risk, Hazards & Crisis in Public Policy*, *11*(2), 116–138. https://doi.org/10.1002/rhc3.12193

Boneham, M. A., & Sixsmith, J. A. (2006). The voices of older women in a disadvantaged community: Issues of health and social capital. *Social Science & Medicine*, *62*(2), 269–279. https://doi.org/10.1016/j.socscimed.2005.06.003

Borgman, C. L. (2015). *Big Data, Little Data, No Data: Scholarship in the Networked World*. The MIT Press.

Bork-Hüffer, T., & Yeoh, B. S. (2017). The geographies of difference in conflating digital and offline spaces of encounter: Migrant professionals' throwntogetherness in Singapore. *Geoforum*, *86*, 93–102. https://doi.org/10.1016/j.geoforum.2017.09.002

Borsboom, D. (2008). Latent Variable Theory. *Measurement: Interdisciplinary Research & Perspective*, *6*(1–2), 25–53. https://doi.org/10.1080/15366360802035497

Bourdieu, P. (2008). The Forms of Capital. In N. W. Biggart (Ed.), *Readings in Economic Sociology* (pp. 280–291). Blackwell Publishers Ltd. http://onlinelibrary.wiley.com/doi/10.1002/9780470755679.ch15/summary

Bourdieu, P., & Wacquant, L. (1992). *An Invitation to Reflexive Sociology*. University Of Chicago Press. http://www.press.uchicago.edu/ucp/books/book/chicago/I/bo3649674.html

Bourgeois, M., & Friedkin, N. E. (2001). The distant core: Social solidarity, social distance and interpersonal ties in core–periphery structures. *Social Networks*, *23*(4), 245–260. https://doi.org/10.1016/S0378-8733(01)00033-8

Boxman, E. A. W., De Graaf, P. M., & Flap, H. D. (1991). The impact of social and human capital on the income attainment of Dutch managers. *Social Networks*, *13*(1), 51–73. https://doi.org/10.1016/0378-8733(91)90013-J

Boyd, E., & Folke, C. (2011). *Adapting Institutions: Governance, Complexity and Social-Ecological Resilience* (1st ed.). Cambridge University Press.

Brandes, U. (2001). A Faster Algorithm for Betweenness Centrality. *Journal of Mathematical Sociology*, *25*(2), 163–177.

Brehm, J., & Rahn, W. (1997). Individual-level evidence for the causes and consequences of social capital. *American Journal of Political Science*, *41*(3), 999–1023. https://doi.org/10.2307/2111684

Breiman, L. (2001a). Random Forests. *Machine Learning*, *45*(1), 5–32.

Breiman, L. (2001b). Statistical Modeling: The Two Cultures (with comments and a rejoinder by the author). *Statistical Science*, *16*(3), 199–231. https://doi.org/10.1214/ss/1009213726

Brockmann, D., Hufnagel, L., & Geisel, T. (2006). The scaling laws of human travel. *Nature*, *439*(7075), 462–465. https://doi.org/10.1038/nature04292

Brody, B. A. (Ed.). (1970). *Readings in the Philosophy of Science*. Prentice Hall.

Brondizio, E. S., Ostrom, E., & Young, O. R. (2009). Connectivity and the Governance of Multilevel Social-Ecological Systems: The Role of Social Capital. *Annual Review of Environment and Resources*, *34*, 253–278. https://doi.org/10.1146/annurev.environ.020708.100707

Browning, C. R., Feinberg, S. L., & Dietz, R. D. (2004). The paradox of social organization: Networks, collective efficacy, and violent crime in urban neighborhoods. *Social Forces*, *83*(2), 503–534. https://doi.org/10.1353/sof.2005.0006

Bryant, C.-A., & Norris, D. A. (2002). Measurement of Social Capital: The Canadian Experience. *Country Report*, 15. http://www.oecd.org/dataoecd/21/48/2381103.pdf

Burdett, R., & Sudjic, D. (Eds.). (2011). *Living in the Endless City: The Urban Age Project by the London School of Economics and Deutsche Bank's Alfred Herrhausen Society*. Phaidon Press.

Burt, R. S. (1997). The Contingent Value of Social Capital. *Administrative Science Quarterly*, *42*(2), 339–365. https://doi.org/10.2307/2393923

Byrne, D. (1998). *Complexity Theory and the Social Sciences: An Introduction* (1st ed.). Routledge.

Cain, N. L., & Nelson, H. T. (2013). What drives opposition to high-voltage transmission lines? *Land Use Policy*, *33*, 204–213. https://doi.org/10.1016/j.landusepol.2013.01.003

Campbell, S. D. (2013). Sustainable Development and Social Justice: Conflicting Urgencies and the Search for Common Ground in Urban and Regional Planning. *Michigan Journal of Sustainability*, *1*. http://dx.doi.org/10.3998/mjs.12333712.0001.007

Canadian Institute of Planners. (2012, December 6). What is Planning? [Official Website]. *Canadian Insititute of Planners*. http://www.cip-icu.ca/web/la/en/pa/3FC2AFA9F72245C4B8D2E709990D58C3/template.asp

Caro, D. H., Sandoval-Hernandez, A., & Luedtke, O. (2014). Cultural, social, and economic capital constructs in international assessments: An evaluation using exploratory structural equation modeling. *School Effectiveness and School Improvement*, *25*(3), 433–450. https://doi.org/10.1080/09243453.2013.812568

Carpiano, R. M. (2006). Toward a neighborhood resource-based theory of social capital for health: Can Bourdieu and sociology help? *Social Science & Medicine*, *62*(1), 165–175. https://doi.org/10.1016/j.socscimed.2005.05.020

Carpiano, Richard M., & Fitterer, L. M. (2014). Questions of trust in health research on social capital: What aspects of personal network social capital do they measure? *Social Science & Medicine*, *116*, 225–234. https://doi.org/10.1016/j.socscimed.2014.03.017

Carta, S. (2020). *Big Data, Code and the Discrete City: Shaping Public Realms*. Routledge.

Castells, M. (2009). *The Rise of the Network Society: The Information Age: Economy, Society, and Culture Volume I* (2nd ed.). Wiley-Blackwell.

Cattell, R. B. (1966). The scree test for the number of factors. *Multivariate Behavioral Research*, *1*, 245–276.

Cayley, D. (Ed.). (2009). *Ideas on the Nature of Science*. Goose Lane Editions.

Chambers, J. M., & Hastie, T. J. (1991). *Statistical Models in S*. Taylor & Francis.

Chen, J.-S., Ching, R. K. H., & Lin, Y.-S. (2004). An Extended Study of the K-Means Algorithm for Data Clustering and Its Applications. *The Journal of the Operational Research Society*, *55*(9), 976–987. https://doi.org/10.2307/4101933

Chen, X., Stanton, B., Gong, J., Fang, X., & Li, X. (2009). Personal Social Capital Scale: An instrument for health and behavioral research. *Health Education Research*, *24*(2), 306–317. https://doi.org/10.1093/her/cyn020

Cheung, C., & Leung, K. (2012). Social Mitigation of the Impact of Urban Renewal on Residents' Morale. *Social Indicators Research*, *106*(3), 523–543. https://doi.org/10.2307/41427011

Chilvers, R., Corr, S., & Singlehurst, H. (2010). Investigation into the occupational lives of healthy older people through their use of time. *Australian Occupational Therapy Journal*, *57*(1), 24–33. https://doi.org/10.1111/j.1440-1630.2009.00845.x

Chomsky, N. (1996). Language and the Problems of Knowledge. In A. P. Martinich (Ed.), *The Philosophy of Language* (3rd ed., pp. 558–557). Oxford University Press.

Chow, W. S., & Chan, L. S. (2008). Social network, social trust and shared goals in organizational knowledge sharing. *Information & Management*, *45*(7), 458–465. https://doi.org/10.1016/j.im.2008-06.007

Christensen, C. (2011). *The Innovator's Dilemma: The Revolutionary Book That Will Change the Way You Do Business* (Reprint). Harper Business.

Christensen, C. M., Baumann, H., Ruggles, R., & Sadtler, T. M. (2006). Disruptive Innovation for Social Change. *Harvard Business Review*, *84*(12), 94–101.

Christensen, G., & Miguel, E. (2018). Transparency, Reproducibility, and the Credibility of Economics Research. *Journal of Economic Literature*, *56*(3), 920–980. https://doi.org/10.1257/jel.20171350

Christensen, R., Johnson, W., Branscum, A., & Hanson, T. E. (2010). *Bayesian Ideas and Data Analysis: An Introduction for Scientists and Statisticians* (1 edition). CRC Press.

Christou, G. G. (2013). *Ariadne's Thread: A Letter to Descartes* [Masters of Architecture]. University of Waterloo.

Chung, E. H., & Shalaby, A. (2005). A trip reconstruction tool for GPS-based personal travel surveys. *Transportation Planning and Technology*, *28*(5), 381–401. https://doi.org/10.1080/03081060500322599

Chung, S., Choi, H., & Lee, S. S. (2014). Measuring Social Capital in the Republic of Korea with Mixed Methods: Application of Factor Analysis and Fuzzy-Set Ideal Type Approach. *Social Indicators Research*, *117*(1), 45–64. https://doi.org/10.1007/s11205-013-0341-8

Cohen, N. (2011, March 26). Cellphones Track Your Every Move, and You May Not Even Know. *The New York Times*. http://www.nytimes.com/2011/03/26/business/media/26privacy.html

Coleman, J. (1994). *Foundations of Social Theory*. Belnap Press.

Coleman, J. S. (1986). Social Theory, Social Research, and a Theory of Action. *American Journal of Sociology*, *91*(6), 1309–1335.

Coleman, J. S. (1988). Social Capital in the Creation of Human Capital. *American Journal of Sociology*, *94*, S95–S120.

Cologna, V., & Siegrist, M. (2020). The role of trust for climate change mitigation and adaptation behaviour: A meta-analysis. *Journal of Environmental Psychology*, *69*, 101428. https://doi.org/10.1016/j.jenvp.2020.101428

Cordeiro, G., M., & De Paula Ferrari, S. L. (1991). A modified score test statistic having chi-squared distribution to order n −1. *Biometrika*, *78*(3), 573–582. https://doi.org/10.1093/biomet/78.3.573

Costello, A. B., & Osborne, J. (2005). Best practices in Exploratory Factor Analysis: Four Recommendations for Getting the Most From Your Analysis. *Practical Assessment Research and Evaluation*, *10*(7). http://pareonline.net/getvn.asp?v=10&n=7

Crandall, D. J., Backstrom, L., Cosley, D., Suri, S., Huttenlocher, D., & Kleinberg, J. (2010). Inferring social ties from geographic coincidences. *Proceedings of the National Academy of Sciences*, *107*(52), 22436–22441. https://doi.org/10.1073/pnas.1006155107

Creswell, J. W. (2008). *Research Design: Qualitative, Quantitative, and Mixed Methods Appro* (3rd ed.). Sage Publications.

Cushing, J. T., & McMullin (Eds.). (1989). *Philosophical Consequences of Quantum Theory: Reflections on Bell's Theorem*. University of Notre Dame Press.

da Silva, J., & Morera, B. (2014). *City Resilience Framework* (City Resilience Index) [Overview]. Rockefeller Foundation and Arup International Development.

Dandekar, H. C. (2005). Qualitative Methods in Planning Research and Practice. *Journal of Architectural and Planning Research*, *22*(2), 129–137.

Davis, P., Gulden, T., & O'Mahony, A. (2017, October). *Priority Challenges for Social and Behavioral Research and Its Modeling* [Presentation]. Computational Social Sciences Society of the Americas, Santa Fe, New Mexico. https://computationalsocialscience.org/events/css2017/

de Jonge, E., & van der Loo, M. (2013). *An Introduction to Data Cleaning with R* (Discussion Paper 60083 201313-X-10–13; p. 53). Statistics Netherland.

de Vries, J. R., van der Zee, E., Beunen, R., Kat, R., & Feindt, P. H. (2019). Trusting the People and the System. The Interrelation Between Interpersonal and Institutional Trust in Collective Action for Agri-Environmental Management. *Sustainability*, *11*(24), 7022. https://doi.org/10.3390/su11247022

Dean, K., Holst, E., Kreiner, S., Schoenborn, C., & Wilson, R. (1994). Measurement Issues in Research on Social Support and Health. *Journal of Epidemiology and Community Health*, *48*(2), 201–206. https://doi.org/10.1136/jech.48.2.201

DeBats, D. A. (2011). Political Consequences of Spatial Organization: Contrasting Patterns in Two Nineteenth-Century Small Cities. *Social Science History*, *35*(4), 505–541. https://doi.org/10.1017/S0145553200011652

Debnath, L. (2010). A Brief Historical Introduction to Euler's Formula for Polyhedra, Topology, Graph Theory and Networks. *International Journal of Mathematical Education in Science and Technology*, *41*(6), 769–785. https://doi.org/10.1080/00207391003675166

Delhey, J., & Welzel, C. (2012). Genealizing Trust: How Outgroup-Trust Grows Beyond Ingroup-Trust. *World Values Research*, *5*(3), 46–69.

Deng, C., Lin, W., Ye, X., Li, Z., Zhang, Z., & Xu, G. (2018). Social media data as a proxy for hourly fine-scale electric power consumption estimation. *Environment and Planning A: Economy and Space*, *50*(8), 1553–1557. https://doi.org/10.1177/0308518X18786250

Derman, E. (2011). *Models. Behaving. Badly.* Free Press.

Diaconis, P., & Mosteller, F. (1989). Methods for Studying Coincidences. *Journal of the American Statistical*

Association, *84*(408), 853–861. https://doi.org/10.1080/01621459.1989.10478847
Dixon, J., & Frolova, Y. (2011). Existential Poverty: Welfare Dependency, Learned Helplessness and Psychological Capital. *Poverty & Public Policy*, *3*(2), 279–298. https://doi.org/10.2202/1944-2858.1158
Doreen Massey. (1995). Places and Their Pasts. *History Workshop Journal*, *39*, 182–192.
Dugundji, E., Scott, D. M., Carrasco, J. A., & Paez, A. (2012). Urban mobility and social-spatial contact-introduction. *Environment and Planning A*, *44*(5), 1011–1015. https://doi.org/10.1068/a45183
Dunbar, R. I. M. (1993). Coveolution of neocortex size, group size and language in humans. *Behavioural and Brain Sciences*, *16*(4), 681–735.
Dunbar, R. I. M. (2011). Constraints on the evolution of social institutions and their implications for information flow. *Journal of Institutional Economics*, *7*(3), 345–371. https://doi.org/10.1017/S1744137410000366
Dunkelman, M. J. (2014). *The Vanishing Neighbor: The Transformation of American Community* (1 edition). W. W. Norton & Company.
Eagle, N. N. (2005). *Machine Perception and Learning of Complex Social Systems* [Dissertation, MIT]. http://reality.media.mit.edu/pdfs/thesis.pdf
Ebdon, D. (1976). On the Underestimation Inherent in the Commonly-Used Formulæ. *Area*, *8*(3), 165–169. JSTOR.
Ebdon, D. (1991). *Statistics in Geography: A Practical Approach - Revised with 17 Programs* (2 edition). Wiley-Blackwell.
Edwards, R. W. (Ed.). (2004). *Measuring Social Capital: An Australian Framework and Indicators* (p. 154) [Information Paper]. Australian Bureau of Statistics. http://www.ausstats.abs.gov.au/Ausstats/free.nsf/Lookup/13C0688F6B98DD45CA256E360077D526/$File/13780_2004.pdf
Enedy, J. D. (1993). Geography and Math: A Technique for Finding Population Centers. *Journal of Geography*, *92*(1), 23–27. https://doi.org/10.1080/00221349308979122
Eng, P. M., Rimm, E. B., Fitzmaurice, G., & Kawachi, I. (2002). Social Ties and Change in Social Ties in Relation to Subsequent Total and Cause-specific Mortality and Coronary Heart Disease Incidence in Men. *American Journal of Epidemiology*, *155*(8), 700–709. https://doi.org/10.1093/aje/155.8.700
Erturk, M. A., Aydin, M. A., Buyukakkaslar, M. T., & Evirgen, H. (2019). A Survey on LoRaWAN Architecture, Protocol and Technologies. *Future Internet*, *11*(10), 216. https://doi.org/10.3390/fi11100216
Espeland, Wendy N., & Stevens, M. L. (1998). Commensuration as a social process. *Annual Review of Sociology*, *24*, 313–343. https://doi.org/10.1146/annurev.soc.24.1.313
Espeland, Wendy Nelson, & Sauder, M. (2007). Rankings and reactivity: How public measures recreate social worlds. *American Journal of Sociology*, *113*(1), 1–40. https://doi.org/10.1086/517897
Ester, M., Kriegel, H.-P., Sander, J., & Xu, X. (1996). *A density-based algorithm for discovering clusters in large spatial databases with noise*. 226–231.
Eubank, S., Guclu, H., Anil Kumar, V. S., Marathe, M. V., Srinivasan, A., Toroczkai, Z., & Wang, N. (2004). Modelling disease outbreaks in realistic urban social networks. *Nature*, *429*(6988), 180-. Academic OneFile.
Evans, J. A., & Aceves, P. (2016). Machine Translation: Mining Text for Social Theory. *Annual Review of Sociology*, *42*(1), 21–50. https://doi.org/10.1146/annurev-soc-081715-074206
Fainstein, S. S., & Campbell, S. (Eds.). (2011). *Readings in Planning Theory* (3rd ed.). Wiley-Blackwell.
Farber, S., Neutens, T., Carrasco, J.-A., & Rojas, C. (2014). Social interaction potential and the spatial distribution of face-to-face social interactions. *Environment and Planning B: Planning and Design*, *41*(6), 960–976. https://doi.org/10.1068/b120034p
Farr, J. (2004). Social Capital: A Conceptual History. *Political Theory*, *32*(1), 6–33.
Feigenbaum, M. (1983). Universal Behavior in Non-Linear Systems. *Physica D*, *7*(1–3), 16–39. https://doi.org/10.1016/0167-2789(83)90112-4
Feldstein, L. M., & Sander, T. (n.d.). *Social Capital Impact Assessment*. John F. Kennedy School of Government, Harvard University, Saguaro Seminar.

http://www.ksg.harvard.edu/saguaro/pdfs/skimpactassessment.pdf
Filion, P., Shipley, R., & Zeralynne, T. (2007). Works planners read: Findings from a Canadian survey. *Canadian Journal of Urban Research*, *22*(June), 59–61.
Finsveen, E., & van Oorschot, W. (2008). Access to Resources in Networks A Theoretical and Empirical Critique of Networks as a Proxy for Social Capital. *Acta Sociologica*, *51*(4), 293–307. https://doi.org/10.1177/0001699308097375
Forbes, A., & Wainwright, S. P. (2001). On the methodological, theoretical and philosophical context of health inequalities research: A critique. *Social Science & Medicine*, *53*(6), 801–816. https://doi.org/10.1016/S0277-9536(00)00383-X
Forrester, J. W. (2009). *Some Basic Concepts in System Dynamics* [D-4894 Course Content]. https://www.cc.gatech.edu/classes/AY2013/cs7601_spring/papers/Forrester-SystemDynamics.pdf
Forster, L. E., & Stoller, E. P. (1992). The Impact of Social Support on Mortality: A Seven-Year Follow-Up of Older Men and Women. *Journal of Applied Gerontology*, *11*(2), 173–186. https://doi.org/10.1177/073346489201100204
Foster, K. A., Smith, R. J., Bell, B. A., & Shaw, T. C. (2019). Testing the Importance of Geographic Distance for Social Capital Resources. *Urban Affairs Review*, *55*(1), 231–256. https://doi.org/10.1177/1078087417714895
Fowler, J. H., & Christakis, N. A. (2008). Dynamic spread of happiness in a large social network: Longitudinal analysis over 20 years in the Framingham Heart Study. *BMJ*, *337*(dec04 2), a2338–a2338. https://doi.org/10.1136/bmj.a2338
Franke, S. (2005). *Measurement of Social Capital: Reference Document for Public Policy Research, Development, and Evaluation* [Interdepartmental research project reference tool]. Policy Research Institute (now Policy Horizons Canada).
Friedmann, J. (2000). The Good City: In Defense of Utopian Thinking. *International Journal of Urban and Regional Research*, *24*(2), 460–472. https://doi.org/10.1111/1468-2427.00258
Friesen, M. J. (2013a). *Is Social Capital Measurement Still Relevant? A Roundtable Policy and Social Impact Assessment White Paper*. Cardus.
Friesen, M. J. (2013b). Social Infrastructure: Underpinning the Success of Cities. *Municipal World*, *123*(12), 11–14.
Friesen, M. J. (2018a). *Navigating Social Capital Scholarship: A Meta-Review of 9164 Web of Science Sources on Social Capital* (p. 44) [Cardus Research Report]. Cardus. file:///C:/Users/mfriesen/Downloads/2018%20-%2001%20-%20Social%20Capital%20Scholarship.pdf
Friesen, M. J. (2018b). Advancing Social Capital Measurement: Using General Social Survey Cycles to Develop an Efficient Survey Instrument. *The American Sociologist*, 1–22. https://doi.org/10.1007/s12108-018-9380-6
Fujiwara, T., & Kawachi, I. (2014). Are Maternal Social Networks and Perceptions of Trust Associated with Suspected Autism Spectrum Disorder in Offspring? A Population-Based Study in Japan. *Plos One*, *9*(7), e101359. https://doi.org/10.1371/journal.pone.0101359
Fukuyama, F. (1996). *Trust: The Social Virtues and The Creation of Prosperity* (1st Free Press Pbk. Ed edition). Free Press.
Fukuyama, F. (1997). Social capital and the modern capitalist society: Creating a high trust workplace. *Stern Business Magazine*, 4.
Fukuyama, F. (2002). Social Capital and Development: The Coming Agenda. *SAIS Review*, *22*(1), 23–37. https://doi.org/10.1353/sais.2002.0009
Gadamer, H.-G. (1960). *Truth and Method* (J. Weinsheimer & D. G. Marshall, Trans.; Second, Revised Edition). Continuum.
Garcia, F. P., Martinez, L. S., & Radoselovics, J. F. de G. (2008). *Estimation of Social Capital in the World* (BBVA Foundacion Working Papers, p. 82) [Working Papers Series]. BBVA Foundacion.
Gates, S., & Justesen, M. K. (2020). Political Trust, Shocks, and Accountability: Quasi-experimental Evidence from a Rebel Attack. *Journal of Conflict Resolution*, 0022002720906446.

https://doi.org/10.1177/0022002720906446
General Social Survey Cycle 22: Social Networks (2008). (2008). [Study Documentation]. Statistics Canada.
General Social Survey Cycle 27: Social Identity (2013). (2013). [Study Documentation]. Statistics Canada.
General Social Survey—2010 Overview of the Time Use of Canadians (Occassional Report Catalogue no. 89-647-X; p. 31). (2011). Statistics Canada. http://www.statcan.gc.ca/pub/89-647-x/89-647-x2011001-eng.pdf
Gesthuizen, M., van der Meer, T., & Scheepers, P. (2009). Ethnic Diversity and Social Capital in Europe: Tests of Putnam's Thesis in European Countries. *Scandinavian Political Studies*, *32*(2), 121–142. https://doi.org/10.1111/j.1467-9477.2008.00217.x
Gidron, N., & Hall, P. A. (2020). Populism as a Problem of Social Integration. *Comparative Political Studies*, *53*(7), 1027–1059. https://doi.org/10.1177/0010414019879947
Giles, L. C., Glonek, G. F. V., Luszcz, M. A., & Andrews, G. R. (2005). Effect of social networks on 10 year survival in very old Australians: The Australian longitudinal study of aging. *Journal of Epidemiology and Community Health*, *59*(7), 574–579. https://doi.org/10.1136/jech.2004.025429
Goldberg, D. (1989). *Genetic Algorithms in Search, Optimization and Machine Learning*. Addison-Wesley.
Goldstein, J. A., Hazy, J. K., & Silberstang, J. (2009). *Complexity Science and Social Entrepreneurship: Adding Social Value Through Systems Thinking*. Isce Publishing.
Gong, H. (2012). A GPS/GIS method for travel mode detection in New York City. *Computers, Environment and Urban Systems*, *36*(2), 131–139.
González, M. C., Hidalgo, C. A., & Barabási, A.-L. (2008). Understanding individual human mobility patterns. *Nature*, *453*(7196), 779–782. https://doi.org/10.1038/nature06958
Goodchild, M. F. (2007). Citizens as sensors: The world of volunteered geography. *GeoJournal*, *69*(4), 211–221.
Gouverneur, D. (2014). *Planning and Design for Future Informal Settlements: Shaping the Self-Constructed City* (1 edition). Routledge.
Government of Canada, P. H. A. of C. (2005, February 23). *Backgrounder—Severe Acute Respiratory Syndrome (SARS)—Public Health Agency of Canada*. http://www.phac-aspc.gc.ca/sars-sras-gen/backgrounder2-eng.php
Graham, S., & McFarlane, C. (Eds.). (2015). *Infrastructural Lives: Urban Infrastructure in Context* (First). Routledge.
Granovetter, M. (2005). The Impact of Social Structure on Economic Outcomes. *The Journal of Economic Perspectives*, *19*(1), 33–50.
Granovetter, M. S. (1973). The Strength of Weak Ties. *American Journal of Sociology*, *78*(6), 1360–1380.
Gregg, M. (2015). Inside the Data Spectacle. *Television & New Media*, *16*(1), 37–51. https://doi.org/10.1177/1527476414547774
Grimmer, J. (2015). We Are All Social Scientists Now: How Big Data, Machine Learning, and Causal Inference Work Together. *PS: Political Science & Politics*, *48*(1), 80–83. Cambridge Core. https://doi.org/10.1017/S1049096514001784
Gunderson, L. H., & Holling, C. S. (2002). *Panarchy: Understanding transformations in human and natural systems*. Island Press.
Haas, T. (2008). *New Urbanism and Beyond: Designing Cities for the Future*. Rizzoli.
Hagen, R. (2019). Collisions Between Institutional and Populist Risk Imaginaries: The "Dark Side" of Negative Asymmetric Thinking. *Sociological Forum*, *34*(S1), 1235–1250. https://doi.org/10.1111/socf.12547
Halevy, A., Norvig, P., & Pereira, F. (2009). The Unreasonable Effectiveness of Data. *IEEE Intelligent Systems*, *24*(2), 8–12. https://doi.org/10.1109/MIS.2009.36
Hampton, K. N., Livio, O., & Sessions Goulet, L. (2010). The Social Life of Wireless Urban Spaces: Internet Use, Social Networks, and the Public Realm. *Journal of Communication*, *60*(4), 701–722. https://doi.org/10.1111/j.1460-2466.2010.01510.x
Hanifan, L. J. (1916). The Rural School Community Center. *Annals of the American Academy of Political and Social Science*, *67*, 130–138.
Harold, J., & Fong, E. (2018). Mnemonic Institutions and Residential Clustering: Jewish Residential

Patterns in Toronto. *Canadian Review of Sociology/Revue Canadienne de Sociologie*, *55*(2), 257–277. https://doi.org/10.1111/cars.12192

Harrison, J. L., Montgomery, C. A., & Jeanty, P. W. (2019). A spatial, simultaneous model of social capital and poverty. *Journal of Behavioral and Experimental Economics*, *78*, 183–192. https://doi.org/10.1016/j.socec.2018.09.001

Hasanzadeh, K., Kyttä, M., Lilius, J., Ramezani, S., & Rinne, T. (2021). Centricity and multi-locality of activity spaces: The varying ways young and old adults use neighborhoods and extra-neighborhood spaces in Helsinki Metropolitan Area. *Cities*, *110*, 103062. https://doi.org/10.1016/j.cities.2020.103062

Hastie, T., Tibshirani, R., & Friedman, J. (2011). *The Elements of Statistical Learning: Data Mining, Inference, and Prediction, Second Edition* (2nd ed. 2009. Corr. 7th printing 2013 edition). Springer.

Hatt, K. (2008). Considering Complexity: Toward A Strategy for Non-linear Analysis. *Canadian Journal of Sociology*, *34*(2), 313–348.

Hatt, K. (2013). Social Attractors: A Proposal to Enhance “Resilience Thinking” about the Social. *Society & Natural Resources*, *26*(1), 30–43. https://doi.org/10.1080/08941920.2012.695859

Henderson, H. V., & Velleman, P. F. (1981). Building Multiple Regression Models Interactively. *Biometrics*, *37*(2), 391–411. https://doi.org/10.2307/2530428

Hildebrandt, M. (2013). Slaves to Big Data, Or Are We? *IDP. Revista De Interenet, Dorecho Y Politica*, *17*, 7–44. https://doi.org/10.7238/ldp.v0i17.1977

Hillier, B. (2015). *Space is the machine: A configurational theory of architecture*. CreateSpace Independent Publishing Platform.

Hillier, B., & Hanson, J. (1989). *The Social Logic of Space* (1st edition). Cambridge University Press.

Hintikka, J. (1998). What is abduction? The fundamental problem of contemporary epistemology. *Transactions of the Charles S Peirce Society*, *34*(3), 503–533.

Hoaglin, D. C. (2003). John W. Tukey and Data Analysis. *Statistical Science*, *18*(3), 311–318.

Holland, J. (2014). *Complexity: A Very Short Introduction* (1 edition). Oxford University Press.

Hollands, R. G. (2008). Will the real smart city please stand up? Intelligent, progressive or entrepreneurial? *City*, *12*(3), 303–320.

Holt-Lunstad, J., Smith, T. B., & Layton, J. B. (2010). Social Relationships and Mortality Risk: A Meta-analytic Review. *PLoS Med*, *7*(7), e1000316. https://doi.org/10.1371/journal.pmed.1000316

Homer-Dixon, T. (2007). *The Upside of Down: Catastrophe, Creativity and the Renewal of Civilization* (First Printing). Vintage Canada.

Hornsey, M. J., Lobera, J., & Diaz-Catalan, C. (2020). Vaccine hesitancy is strongly associated with distrust of conventional medicine, and only weakly associated with trust in alternative medicine. *Social Science & Medicine*, *255*, 113019. https://doi.org/10.1016/j.socscimed.2020.113019

Horsten, M., Mittleman, M. A., Wamala, S. P., Schenck-Gustafsson, K., & Orth-Gomér, K. (2000). Depressive symptoms and lack of social integration in relation to prognosis of CHD in middle-aged women. The Stockholm Female Coronary Risk Study. *European Heart Journal*, *21*(13), 1072–1080. https://doi.org/10.1053/euhj.1999.2012

Hu, C. (2014). Life and Death in the Open System. *On Site Review*, *32*, 12–14.

Humphreys, P. (2016). *Emergence* (1st ed.). Oxford University Press.

Huss, A., Beekhuizen, J., Kromhout, H., & Vermeulen, R. (2014). Using GPS-derived speed patterns for recognition of transport modes in adults. *International Journal of Health Geographics*, *13*, 40. https://doi.org/10.1186/1476-072X-13-40

Inglehart, R. (1997). *Modernization and Postmodernization: Cultural, Economic, and Political Change in 43 Societies*. Princeton University Press.

Innes, J. E., & Booher, D. E. (2010). *Planning with Complexity: An Introduction to Collaborative Rationality for Public Policy* (1st ed.). Routledge.

Irwin, E. G. (2010). New Directions for Urban Economic Models of Land Use Change: Incorporating Spatial Dynamics and Heterogeneity. *Journal of Regional Science*, *50*(1), 65–91.

Jackson, L. E. (2003). The relationship of urban design to human health and condition. *Landscape and*

Urban Planning, *64*(4), 191–200. https://doi.org/10.1016/S0169-2046(02)00230-X
Jacobs, J. (1959). *The Death and Life of Great American Cities* (1REV ed.). Vintage.
Jahanmiri, F. (2014). *Fractal Application in Urban Planning and Design* [801/802 Ph.D. Seminar].
Janssen, M. A., Holahan, R., Lee, A., & Ostrom, E. (2010). Lab Experiments for the Study of Social-Ecological Systems. *Science*, *328*(5978), 613–617. https://doi.org/10.1126/science.1183532
Jantsch, E. (1972). *Technological Planning and Social Futures*. John Wiley & Sons.
Johnson, R. B., & Onwuegbuzie, A. J. (2004). Mixed Methods Research: A Research Paradigm Whose Time Has Come. *Educational Researcher*, *33*(7), 14–26.
Johnston, E. (2010). Governance Infrastructures in 2020. *Public Administration Review*, *70*, S122–S128.
Jolliffe, I., & Morgan, B. (1992). Principal component analysis and exploratory factor analysis. *Statistical Methods in Medical Research*, *1*(1), 69–95. https://doi.org/10.1177/096228029200100105
Kalkhan, M. A. (2011). *Spatial Statistics: GeoSpatial Information Modeling and Thematic Mapping* (1st ed.). CRC Press.
Kawachi, I., Colditz, G. A., Ascherio, A., Rimm, E. B., Giovannucci, E., Stampfer, M. J., & Willett, W. C. (1996). A prospective study of social networks in relation to total mortality and cardiovascular disease in men in the USA. *Journal of Epidemiology and Community Health*, *50*(3), 245–251.
Ke, Q., Ferrara, E., Radicchi, F., & Flammini, A. (2015). Defining and identifying Sleeping Beauties in science. *Proceedings of the National Academy of Sciences of the United States of America*, *112*(24), 7426–7431. https://doi.org/10.1073/pnas.1424329112
Keuschnigg, M., Lovsjö, N., & Hedström, P. (2018). Analytical sociology and computational social science. *Journal of Computational Social Science*, *1*(1), 3–14. https://doi.org/10.1007/s42001-017-0006-5
Khabsa, M., Elmagarmid, A., Ilyas, I., Hammady, H., & Ouzzani, M. (2016). Learning to identify relevant studies for systematic reviews using random forest and external information. *Machine Learning*, *102*(3), 465–482. https://doi.org/10.1007/s10994-015-5535-7
Knigge, L., & Cope, M. (2006). Grounded Visualization: Integrating the Analysis of Qualitative and Quantitative Data through Grounded Theory and Visualization. *Environment and Planning A: Economy and Space*, *38*(11), 2021–2037. https://doi.org/10.1068/a37327
Knoke, D. (1999). Organizational Networks and Corporate Social Capital. In R. T. A. J. Leenders & S. M. Gabbay (Eds.), *Corporate Social Capital and Liability* (pp. 17–42). Springer US. https://doi.org/10.1007/978-1-4615-5027-3_2
Kojima, K. (2014). Crafting Space: Generative Processes of Architectural Configurations. *Architectural Design*, *Profile No. 231*(Emphatic Space), 38–45.
Kuhn, S., Egert, B., Neumann, S., & Steinbeck, C. (2008). Building blocks for automated elucidation of metabolites: Machine learning methods for NMR prediction. *BMC Bioinformatics*, *9*, 400. https://doi.org/10.1186/1471-2105-9-400
Kuhn, T. (1962). *The Structure of Scientific Revolutions* (3rd ed.). University of Chicago Press.
Kuhn, T. (1970). The Function of Dogma in Scientific Research. In *Readings in the Philosophy of Science* (pp. 356–373). Prentice-Hall.
Kwakkel, J. H., Carley, S., Chase, J., & Cunningham, S. W. (2014). Visualizing geo-spatial data in science, technology and innovation. *Technological Forecasting and Social Change*, *81*(0), 67–81. https://doi.org/10.1016/j.techfore.2012.09.007
Kwan, M.-P. (2007). Mobile communications, social networks, and urban travel: Hypertext as a new metaphor for conceptualizing spatial interaction. *Professional Geographer*, *59*(4), 434–446. https://doi.org/10.1111/j.1467-9272.2007.00633.x
Landy, D., Guay, B., & Marghetis, T. (2018). Bias and ignorance in demographic perception. *Psychonomic Bulletin & Review*, *25*(5), 1606–1618. https://doi.org/10.3758/s13423-017-1360-2
Lauer, S. R., & Yodanis, C. L. (2004). The International Social Survey Programme (ISSP): A Tool for Teaching with an International Perspective. *Teaching Sociology*, *32*(3), 304–313. https://doi.org/10.2307/3211408
Lavric, A. (2019). LoRa (Long-Range) High-Density Sensors for Internet of Things. *Journal of Sensors*, *2019*, 3502987. https://doi.org/10.1155/2019/3502987
Leavelle, T. N. (2004). Geographies of Encounter: Religion and Contested Spaces in Colonial North

America. *American Quarterly*, *56*(4), 913–943. https://doi.org/10.1353/aq.2004.0060

Leek, J. T., & Peng, R. D. (2015). Opinion: Reproducible research can still be wrong: Adopting a prevention approach. *Proceedings of the National Academy of Sciences of the United States of America*, *112*(6), 1645–1646. https://doi.org/10.1073/pnas.1421412111

Lefebvre, H. (1992). *The Production of Space* (D. Nicholson-Smith, Trans.; 1st edition). Wiley-Blackwell.

Lewin, R. (1999). *Complexity: Life at the edge of chaos*. University of Chicago Press.

Lim, Y., Edelenbos, J., & Gianoli, A. (2019). Identifying the results of smart city development: Findings from systematic literature review. *Cities*, *95*. https://doi.org/10.1016/j.cities.2019.102397

Lin, C.-Y., Hsu, C.-Y., Gunnell, D., Chen, Y.-Y., & Chang, S.-S. (2019). Spatial patterning, correlates, and inequality in suicide across 432 neighborhoods in Taipei City, Taiwan. *Social Science & Medicine*, *222*, 20–34. https://doi.org/10.1016/j.socscimed.2018.12.011

Lin, N., Cook, K. S., & Burt, R. S. (2001). *Social capital: Theory and research*. Transaction Publishers.

Lindblom, C. E. (1959). The Science of "Muddling Through." *Public Administration Review*, *19*(2), 79–88. https://doi.org/10.2307/973677

Lindström, C., & Lindström, M. (2006). "Social Capital," GNP Per Capita, Relative Income, and Health: An Ecological Study of 23 Countries. *International Journal of Health Services*, *36*(4), 679–696. https://doi.org/10.2190/C2PP-WF4R-X081-W2QN

Little, D. (2002, September). *Transportation as a Large Historical Factor: A Case Study in New Philosophy of History* [Colloquium]. Humanitas Colloquium, University of Michigan-Dearborn.

Lochner, K., Kawachi, I., & Kennedy, B. P. (1999). Social capital: A guide to its measurement. *Health & Place*, *5*(4), 259–270. https://doi.org/10.1016/S1353-8292(99)00016-7

Logan, J. R. (2012). Making a Place for Space: Spatial Thinking in Social Science. *Annual Review of Sociology*, *38*, 507–524.

Lorenz, E. N. (1963). Deterministic Nonperiodic Flow. *Journal of the Atmospheric Sciences*, *20*(2), 130–141. https://doi.org/10.1175/1520-0469(1963)020<0130:DNF>2.0.CO;2

Loury, G. (1992). The economics of discrimination: Getting to the core of the problem. *Harvard Journal for African American Public Policy*, *1*, 91–110.

Macinko, J., & Starfield, B. (2001). The Utility of Social Capital in Research on Health Determinants. *The Milbank Quarterly*, *79*(3), 387–427. https://doi.org/10.1111/1468-0009.00213

Major, M. D., Studies, U. C., London. Bartlett School of Graduate, & Laboratory, U. C., London. Space Syntax. (1997). *Space syntax: First International Symposium : [proceedings], Space Syntax Laboratory, The Bartlett School of Graduate Studies, University College, London, 16-18 April 1997*. Bartlett School of Graduate Studies, University College, London.

Mandel, J. (1972). Principal components, analysis of variance and data structure. *Statistica Neerlandica*, *26*(3), 119–129. https://doi.org/10.1111/j.1467-9574.1972.tb00180.x

March, L., & Steadman, P. (1974). *The geometry of environment: An introduction to spatial organization in design*. M.I.T. Press.

Marche, S. (2012, May). Is Facebook Making Us Lonely? *The Atlantic*. http://www.theatlantic.com/magazine/archive/2012/05/is-facebook-making-us-lonely/8930/

Mardia, K. V., & Jupp, P. E. (1999). *Directional Statistics* (2 edition). Wiley.

Margolis, M. (2012). *Arduino Cookbook* (2nd ed.). O'Reilly.

Mastrandrea, R., Fournet, J., & Barrat, A. (2015). Contact patterns in a high school: A comparison between data collected using wearable sensors, contact diaries and friendship surveys. *ArXiv:1506.03645 [Physics]*. http://arxiv.org/abs/1506.03645

Mathias, A., Páez, D., Espinosa, A., Sandoval, S., Alzugaray, C., Arnoso, M., Cárdenas, M., Costa, S. da, Reyes, C., Rimé, B., & Zubieta, E. (2020). The association between Truth Commissions evaluation, emotional climate and institutional trust: Comparison and meta-analysis of surveys in six South American countries (La relación entre la evaluación de las Comisiones de la verdad, el clima emocional y la confianza institucional: comparación y metanálisis en seis países de Sudamérica). *International Journal of Social Psychology*, *35*(2), 203–245. https://doi.org/10.1080/02134748.2020.1721053

Mayer-Schonberger, V., & Cukier, K. (2013). *Big Data: A Revolution that wil Transform How We Live, Work*

and Think. Houghton Mifflin.
McDoom, O. S. (2014). Predicting violence within genocide: A model of elite competition and ethnic segregation from Rwanda. *Political Geography*, *42*, 34–45. https://doi.org/10.1016/j.polgeo.2014.05.006
McPherson, M., Smith-Lovin, L., & Brashears, M. E. (2006). Social Isolation in America: Changes in Core Discussion Networks over Two Decades. *American Sociological Review*, *71*(3), 353–375.
McShane, B. B., Gal, D., Gelman, A., Robert, C., & Tackett, J. L. (2017). Abandon Statistical Significance. *ArXiv:1709.07588 [Stat]*. http://arxiv.org/abs/1709.07588
Mehaffy, M. W., & Alexander, C. (2016). *A City is Not a Tree: 50th Anniversary Edition*. Sustasis Press.
Mellinger, M. (1987). Multivariate data analysis. *Chemometrics and Intelligent Laboratory Systems*, *2*(1), 29–36. https://doi.org/10.1016/0169-7439(87)80083-7
Miller, H. J. (2004). Tobler's First Law and Spatial Analysis. *Annals of the Association of American Geographers*, *94*(2), 284–289.
Mingo, I., & Faggiano, M. P. (2020). Trust in Institutions Between Objective and Subjective Determinants: A Multilevel Analysis in European Countries. *Social Indicators Research: An International and Interdisciplinary Journal for Quality-of-Life Measurement*, 1–25.
Mir, R., & Watson, A. (2001). Critical Realism and Constructivism in Strategy Research: Toward a Synthesis. *Strategic Management Journal*, *22*(12), 1169–1173.
Mitchell, A. (1999). *The Esri Guide to GIS Analysis, Volume 1: Geographic Patterns and Relationships* (1 edition). Esri Press.
Mitchell, A. (2005). *The Esri Guide to GIS Analysis, Volume 2: Spatial Measurements and Statistics* (1 edition). Esri Press.
Moguerza, J. M., & Muñoz, A. (2006). Support Vector Machines with Applications. *Statistical Science*, *21*(3), 322–336.
Mohan, J., Twigg, L., Barnard, S., & Jones, K. (2005). Social capital, geography and health: A small-area analysis for England. *Social Science & Medicine*, *60*(6), 1267–1283. https://doi.org/10.1016/j.socscimed.2004.06.050
Montgomery, C. (2014, October). Charles Montgomery: Welcome to Happyville: How we can improve our lives by designing smart cities. *National Post*. http://fullcomment.nationalpost.com/2014/10/11/charles-montgomery-welcome-to-happyville-how-we-can-improve-our-lives-by-designing-smart-cities/
Morgan, J. N., & Sonquist, J. A. (1963). Problems in the Analysis of Survey Data and a Proposal. *Journal of the American Statistical Association*, *58*, 414–435.
Mosier, K. L., & Fischer, U. M. (2010). *Informed by Knowledge: Expert Performance in Complex Situations* (1st ed.). Psychology Press.
Myint, L., Leek, J. T., & Jager, L. R. (2017). Explanation implies causation? *BioRxiv*, 218784. https://doi.org/10.1101/218784
Nahapiet, J., & Ghoshal, S. (1998). Social Capital, Intellectual Capital, and the Organizational Advantage. *The Academy of Management Review*, *23*(2), 242–266. https://doi.org/10.2307/259373
Navarrete, R. M., & Castillo-Ortiz, P. (2020). Constitutional courts and citizens' perceptions of judicial systems in Europe. *Comparative European Politics*, *18*(2), 128–150. https://doi.org/10.1057/s41295-019-00154-9
Neutens, T., Farber, S., Delafontaine, M., & Boussauw, K. (2013). Spatial variation in the potential for social interaction: A case study in Flanders (Belgium). *Computers Environment and Urban Systems*, *41*, 318–331. https://doi.org/10.1016/j.compenvurbsys.2012.06.007
Neville, P. (1999). Growing Trees for Stratified Modeling. *Computing Science and Statistics*, *30*.
Newman, M. E. J. (2010). *Networks: An Introduction*. Oxford University Press.
Nguyen, L., Szkudlarek, B., & Seymour, R. G. (2015). Social impact measurement in social enterprises: An interdependence perspective. *Canadian Journal of Administrative Sciences / Revue Canadienne Des Sciences de l'Administration*, *32*(4), 224–237. https://doi.org/10.1002/cjas.1359
Nguyen, T., Andreasen, S. Z., Wolff, A., & Bang, D. D. (2018). From Lab on a Chip to Point of Care Devices: The Role of Open Source Microcontrollers. *Micromachines*, *9*(8), 403.

Nisbet, R. (1953). *The Quest for Community: A Study in the Ethics of Order and Freedom*. Intercollegiate Studies Institute.

Omrani, H. (2015). Predicting Travel Mode of Individuals by Machine Learning. *Transportation Research Procedia*, *10*(Complete), 840–849. https://doi.org/10.1016/j.trpro.2015.09.037

Omrani, H., Charif, O., Gerber, P., Awasthi, A., & Trigano, P. (2013). Prediction of Individual Travel Mode with Evidential Neural Network Model. *Transportation Research Record: Journal of the Transportation Research Board*, *2399*, 1–8. https://doi.org/10.3141/2399-01

O'Neil, C. (2016). *Weapons of Math Destruction: How Big Data Increases Inequality and Threatens Democracy*. Crown.

Onnela, J.-P., Arbesman, S., González, M. C., Barabási, A.-L., & Christakis, N. A. (2011). Geographic Constraints on Social Network Groups. *PLoS ONE*, *6*(4), e16939. https://doi.org/10.1371/journal.pone.0016939

Orban, K., Edberg, A.-K., & Erlandsson, L.-K. (2012). Using a time-geographical diary method in order to facilitate reflections on changes in patterns of daily occupations. *Scandinavian Journal of Occupational Therapy*, *19*(3), 249–259. https://doi.org/10.3109/11038128.2011.620981

Ostrom, E. (1986). An Agenda for the Study of Institutions. *Public Choice*, *48*(1), 3–25.

Ostrom, E. (2005). *Understanding institutional diversity*. Princeton University Press.

Ostrom, E., & Ahn, T. K. (2003). *Foundations of Social Capital*. Edward Elgar Pub.

Pacheco, P. J. (2018). After Planning, the Production of Radical Social Space in Barcelona: Real-Estate Financial Circuit and (De Facto) Right to the City. *Urban Planning*, *3*(3), 83–105. https://doi.org/10.17645/up.v3i3.1360

Page, S. E. (2017, October). *Many Models Thinking* [Presentation]. Computational Social Sciences Society of the Americas, Santa Fe, New Mexico. https://computationalsocialscience.org/events/css2017/

Paldam, M. (2000). Social capital: One or many? Definition and measurement. *Journal of Economic Surveys*, *14*(5), 629–653. https://doi.org/10.1111/1467-6419.00127

Pantell, M., Rehkopf, D., Jutte, D., Syme, S. L., Balmes, J., & Adler, N. (2013). Social Isolation: A Predictor of Mortality Comparable to Traditional Clinical Risk Factors. *American Journal of Public Health*, *103*(11), 2056–2062. bth.

Papinski, Dominik, Scott, D. M., & Doherty, S. T. (2009). Exploring the route choice decision-making process: A comparison of planned and observed routes obtained using person-based GPS. *Transportation Research Part F-Traffic Psychology and Behaviour*, *12*(4), 347–358. https://doi.org/10.1016/j.trf.2009.04.001

Papinski, Dorninik, & Scott, D. M. (2011). A GIS-based toolkit for route choice analysis. *Journal of Transport Geography*, *19*(3), 434–442. https://doi.org/10.1016/j.jtrangeo.2010.09.009

Patil, P., Peng, R. D., & Leek, J. T. (2016). What Should Researchers Expect When They Replicate Studies? A Statistical View of Replicability in Psychological Science. *Perspectives on Psychological Science*, *11*(4), 539–544. https://doi.org/10.1177/1745691616646366

Pavlovskaya, M. (2006). Theorizing with GIS: A Tool for Critical Geographies? *Environment and Planning A: Economy and Space*, *38*(11), 2003–2020. https://doi.org/10.1068/a37326

Pearce, N., & Davey Smith, G. (2003). Is Social Capital the Key to Inequalities in Health? *American Journal of Public Health*, *93*(1), 122–129. https://doi.org/10.2105/AJPH.93.1.122

Pennar, K. (1997, December 15). The ties that lead to prosperity: The economic value of social bonds is only beginning to be measured. *Business Week*, 153–153.

Pentland, A. (2014). *Social physics: How ideas turn into action.* Penguin Books.

Petäjäjärvi, J., Mikhaylov, K., Pettissalo, M., Janhunen, J., & Iinatti, J. (2017). Performance of a low-power wide-area network based on LoRa technology: Doppler robustness, scalability, and coverage. *International Journal of Distributed Sensor Networks*, *13*(3), 155014771769941-. https://doi.org/10.1177/1550147717699412

Peterson, M. (2016). Living with difference in hyper-diverse areas: How important are encounters in semi-public spaces? *Social & Cultural Geography*, *18*(8), 1067–1085. https://doi.org/10.1080/14649365.2016.1210667

Petroni, A., Cuomo, F., Schepis, L., Biagi, M., Listanti, M., & Scarano, G. (2018). Adaptive Data

Synchronization Algorithm for IoT-Oriented Low-Power Wide-Area Networks. *Sensors*, *18*(11), 4053. https://doi.org/10.3390/s18114053

Pflieger, G. (2008). *The social fabric of the networked city*. EPFL Press.

Piekut, A., & Valentine, G. (2017). Spaces of encounter and attitudes towards difference: A comparative study of two European cities. *Social Science Research*, *62*, 175–188. https://doi.org/10.1016/j.ssresearch.2016.08.005

Pinker, S. (2014). *The Village Effect: How Face-to-Face Contact Can Make Us Healthier, Happier, and Smarter*. Spiegel & Grau.

Portes, A. (1998). Social Capital: Its origins and applications in modern sociology. *Annual Review of Sociology*, *24*, 1–24. https://doi.org/10.1146/annurev.soc.24.1.1

Portes, Alejandro, & Sensenbrenner, J. (1993). Embeddedness and Immigration: Notes on the Social Determinants of Economic Action. *American Journal of Sociology*, *98*(6), 1320–1350.

Postman, N. (1993). *Technopoly: The Surrender of Culture to Technology* (Reprint edition). Vintage.

Putnam, H. (1988). *Representation and Reality* (6th Printing). The MIT Press.

Putnam, R. D. (1995). Bowling Alone—Revisted. *The Responsive Community*, *5*(2), 18–33.

Putnam, R. D. (2001). *Bowling Alone*. Simon and Schuster.

Putnam, R. D., Leonardi, R., & Nanetti, R. Y. (1994). *Making Democracy Work: Civic Traditions in Modern Italy*. Princeton University Press.

Qadir, Q. M., Rashid, T. A., Al-Salihi, N. K., Ismael, B., Kist, A. A., & Zhang, Z. (2018). Low Power Wide Area Networks: A Survey of Enabling Technologies, Applications and Interoperability Needs. *IEEE Access*, *6*, 77454–77473. https://doi.org/10.1109/ACCESS.2018.2883151

Quine, W. V. O. (1960). *Word and Object*. The MIT Press.

R Development Core Team. (2016). *R: A Language and Environment for Statistical Computing* (3.2.5) [X86_64-w64-mingw32/x64 (64-bit)]. R Foundation for Statistical Computing.

Raan, A. F. J. van. (2004). Sleeping Beauties in science. *Scientometrics*, *59*(3), 467–472. https://doi.org/10.1023/B:SCIE.0000018543.82441.f1

Reese, L. A., & Rosenfeld, R. (Eds.). (2012). *Comparative Civic Culture: The Role of Local Culture in Urban Policy-Making*. Ashgate Publishing, Ltd.

Reinikainen, H., Kari, J. T., & Luoma-aho, V. (2020). Generation Z and Organizational Listening on Social Media. *Media and Communication*, *8*(2), 185–196. https://doi.org/10.17645/mac.v8i2.2772

Richardson, K. A. (2010). *Thinking About Complexity: Grasping the Continuum through Criticism and Pluralism*. Emergent Publications.

Richardson, K. A., Snowden, D., & Allen, P. M. (2006). *Emergence: Complexity & Organization 2005 Annual*. Isce Publishing.

Richey, S. (2007). Manufacturing Trust: Community Currencies and the Creation of Social Capital. *Political Behavior*, *29*(1), 69–88. https://doi.org/10.2307/4500234

Ried, K., Müller, T., & Briegel, H. J. (2019). Modelling collective motion based on the principle of agency: General framework and the case of marching locusts. *PLOS ONE*, *14*(2), e0212044. https://doi.org/10.1371/journal.pone.0212044

Robinson, J. P. (1999). The time-diary method—Structure and uses. In W. E. Pentland, A. S. Harvey, M. P. Lawton, & M. A. McColl (Eds.), *Time Use Research in the Social Sciences* (pp. 47–89). Kluwer Academic/Plenum Publ.

Rohac, D., Kumar, S., & Johansson Heinö, A. (2017). The wisdom of demagogues: Institutions, corruption and support for authoritarian populists. *Economic Affairs (Harlow)*, *37*(3), 382–396. https://doi.org/10.1111/ecaf.12264

Roo, G. D., & Silva, E. A. (Eds.). (2010). *A Planner's Encounter With Complexity*. Ashgate Pub Co.

Rooduijn, M. (2014). The Nucleus of Populism: In Search of the Lowest Common Denominator. *Government and Opposition*, *49*(4), 573–599. http://dx.doi.org.proxy.lib.uwaterloo.ca/10.1017/gov.2013.30

Rorty, R. (1979). *Philosophy and the Mirror of Nature*. Princeton University Press.

Rorty, R. (1998). *Truth and Progress* (Vol. 1). Cambridge University Press.

Rosa, T. R., Betim, F. S., & Ferreira, R. de Q. (2017). Development and application of a labmade apparatus using open-source "arduino" hardware for the electrochemical pretreatment of boron-doped diamond electrodes. *Electrochimica Acta*, *231*, 185–189. https://doi.org/10.1016/j.electacta.2017.01.180

Rostila, M. (2011). The Facets of Social Capital. *Journal for the Theory of Social Behaviour*, *41*(3), 308–326. https://doi.org/10.1111/j.1468-5914.2010.00454.x

Rovisco, M. (2010). Reframing Europe and the Global: Conceptualizing the Border in Cultural Encounters. *Environment and Planning. D, Society & Space*, *28*(6), 1015–1030. https://doi.org/10.1068/d0809

Rupasingha, A., Goetz, S. J., & Freshwater, D. (2002). Social and institutional factors as determinants of economic growth: Evidence from the United States counties. *Papers in Regional Science*, *81*(2), 139–155. https://doi.org/10.1007/s101100100091

Salzberg, S. L. (1994). Book Review: C4.5 Programs for Machine Learning by J. Ross Quinlan. *Machine Learning*, *16*(3), 235–240. https://doi.org/10.1007/BF00993309

Sampson, R. J., & Graif, C. (2009). Neighborhood Social Capital as Differential Social Organization Resident and Leadership Dimensions. *American Behavioral Scientist*, *52*(11), 1579–1605. https://doi.org/10.1177/0002764209331527

Sampson, R. J., & Raudenbush, S. W. (1999). Systematic Social Observation of Public Spaces: A New Look at Disorder in Urban Neighborhoods. *American Journal of Sociology*, *105*(3), 603–651. https://doi.org/10.1086/210356

Sancton, A. (2000). Jane Jacobs on the organization of municipal government. *Journal of Urban Affairs*, *22*(4), 463–471. https://doi.org/10.1111/0735-2166.00067

Sanders, T. I. (1998). *Strategic thinking and the new science: Planning in the midst of chaos, complexity, and change*. Free Press.

Savitz, N. V., & Raudenbush, S. W. (2009). Exploiting Spatial Dependence to Improve Measurement of Neighborhood Social Processes. *Sociological Methodology*, *39*, 151–183.

Schelling, T. (2006). *Micromotives and Macrobehavior* (Revised edition). W. W. Norton & Company.

Schiff, M. (1992). Social Capital, Labor Mobility, and Welfare: The Impact of Uniting States. *Rationality and Society*, *4*(2), 157–175. https://doi.org/10.1177/1043463192004002003

Schoenbach, V. J., Kaplan, B. H., Fredman, L., & Kleinbaum, D. G. (1986). Social Ties and Mortality in Evans County, Georgia. *American Journal of Epidemiology*, *123*(4), 577–591.

Schumpeter. (2011, January 29). The Crescent and the Company. *The Economist, January 2011*, 67.

Schurmann, C., Moeckel, R., & Wegener, M. (2002). *Microsimulation of urban land use*. European Regional Science Association. http://EconPapers.repec.org/RePEc:wiw:wiwrsa:ersa02p261

Scott, D. M., Dugundji, E. R., & Paez, A. (2013). The social dimension of activity, travel and location choice behavior. *Journal of Transport Geography*, *31*, 212–215. https://doi.org/10.1016/j.jtrangeo.2013.06.009

Scott, L., & Janikas, M. V. (2010). Spatial Statistics in ArcGIS. In *Handbook of Applied Spatial Statistics: Software Tools, Methods and Applications (M. M. Fischer and A. Getis, eds.)* (pp. 27–41). Springer-Verlag.

Sekar, A., & Rai, V. (2018). Inferring Latent Constructs from Passive Datasets: Significance and Opportunities. *SSRN Electronic Journal*, 1–19. https://doi.org/DOI: 10.2139/ssrn.3285417

Sekara, V., Stopczynski, A., & Lehmann, S. (2016). Fundamental structures of dynamic social networks. *Proceedings of the National Academy of Sciences of the United States of America*, *113*(36), 9977–9982. https://doi.org/10.1073/pnas.1602803113

Sennett, R. (2012). *Together: The rituals, pleasures and politics of cooperation*. Allen Lane.

Serre, D., Barroca, B., & Laganier, R. (Eds.). (2012). *Resilience and Urban Risk Management* (Har/Cdr edition). CRC Press.

Serres, M. (1995). *Conversations on Science, Culture, and Time: Michel Serres with Bruno LaTour*. University of Michigan Press.

Shoval, N., & Isaacson, M. (2009). *Tourist Mobility and Advanced Tracking Technologies* (1st ed.). Routledge.

Simon, H. A. (1962). The Architecture of Complexity. *Proceedings of the American Philosophical Society*, *106*(6), 467–482.

Sivram Kish. (2011, December 10). *Open Generative Design* [About]. http://www.opengenerativedesign.com/
Small, Mario L., & Adler, L. (2019). The Role of Space in the Formation of Social Ties. *The Annual Review of Sociology*, *45*, 111–132. https://doi.org/10.1146/annurev-soc-073018- 022707
Small, Mario Luis. (2011). How to Conduct a Mixed Methods Study: Recent Trends in a Rapidly Growing Literature. *Annual Review of Sociology*, *37*, 57–86. https://doi.org/10.2307/41288599
Smith, E. R., & Moreno, J. L. (1951). Sociometry, Experimental Method and the Science of Society. *The American Catholic Sociological Review*, *12*(3), 176. https://doi.org/10.2307/3706642
Snowden, D., & Boone, M. E. (2007). Leader's Framework for Decision Making. *Harvard Business Review*, *November*, 10.
Staff Writer. (1936, July 15). Realty Boom. *Hamilton Spectator*.
Stanley, M. (2007). So simple a thing as a star: The Eddington–Jeans debate over astrophysical phenomenology. *The British Journal for the History of Science*, *40*(1), 53–82. https://doi.org/10.1017/S0007087406008740
Steinmetz, G. (1998). Critical Realism and Historical Sociology. A Review Article. *Comparative Studies in Society and History*, *40*(1), 170–186.
Sugrue, T. J. (2005). *The Origins of the Urban Crisis: Race and Inequality in Postwar Detroit* (With a New preface by the author edition). Princeton University Press.
Suhr, D. D. (2005, April 10). Statistics and Data Analysis Paper 203-30 Principal Component Analysis vs. Exploratory Factor Analysis. *Thirtieth Annual*. SAS Users Group International Conference (SUGI 30), Philadelphia.
Syme, S. L. (1991). Control and health: A personal perspective. *Advances*, *7*(2), 16–27.
Takahashi, B., & Tandoc, E. C. (2016). Media sources, credibility, and perceptions of science: Learning about how people learn about science. *Public Understanding of Science*, *25*(6), 674–690. https://doi.org/10.1177/0963662515574986
Taleb, N. N. (2010). *The Black Swan: Second Edition: The Impact of the Highly Improbable: With a new section: "On Robustness and Fragility"* (2nd ed.). Random House Trade Paperbacks.
Talen, E. (1999). Sense of community and neighbourhood form: An assessment of the social doctrine of new urbanism. *Urban Studies*, *36*(8), 1361–1379. https://doi.org/10.1080/0042098993033
Ten Brinke, L., & Keltner, D. (2020). Theories of power: Perceived strategies for gaining and maintaining power. *Journal of Personality and Social Psychology*. https://doi.org/10.1037/pspi0000345
Thomas, C. Y. (1996). Capital Markets, Financial Markets, and Social Capital (An Essay on Economic Theory and Economic Ideas). *Social and Economic Studies*, *45*(2/3), 1–23.
Thomas, G. (1997). What's the use of theory? *Harvard Educational Review*, *67*(1), 75–104.
Thomas, M., Hunt, A., Hurley, M., Robertson, S., & Carter, B. (2011). Time-use diaries are acceptable to parents with a disabled preschool child and are helpful in understanding families' daily lives. *Child Care Health and Development*, *37*(2), 168–174. https://doi.org/10.1111/j.1365-2214.2010.01156.x
Thompson, J., McClure, R., & Desilva, A. (2017, October). *A Complex Systems Approach for Understanding the Effect of Policy and Management Interventions on Health System Performance* [Presentation]. Computational Social Sciences Society of the Americas, Santa Fe, New Mexico. https://computationalsocialscience.org/events/css2017/
Thompson, M., & Oelker, A. (2013). Use of Participant-Generated Photographs Versus Time Use Diaries as a Method of Qualitative Data Collection. *International Journal of Qualitative Methods*, *12*, 624–637. https://doi.org/10.1177/160940691301200133
Thrift, Nigel. (2005). But malice aforethought: Cities and the natural history of hatred. *Transactions - Institute of British Geographers (1965)*, *30*(2), 133–150. https://doi.org/10.1111/j.1475-5661.2005.00157.x
Thrift, Nj. (1983). On the Determination of Social-Action in Space and Time. *Environment and Planning D-Society & Space*, *1*(1), 23–57. https://doi.org/10.1068/d010023
Tocqueville, A. de. (2001). *Democracy in America* (R. C. Heffner, Ed.). Signet Classics.
Tod, A. M., Kennedy, F., Stocks, A.-J., McDonnell, A., Ramaswamy, B., Wood, B., & Whitfield, M. (2016).

Good-quality social care for people with Parkinson's disease: A qualitative study. *BMJ Open*, *6*(2), e006813. https://doi.org/10.1136/bmjopen-2014-006813

Trofimova, I. (2000). Principles, concepts and phenomena of Ensembles with Variable Structure. In *Nonlinear Dynamics in Life and Social Sciences*. IOS Press.

Tukey, J. W. (1962). The Future of Data Analysis. *The Annals of Mathematical Statistics*, *33*(1), 1–67.

Uphoff, N., & Wijayaratna, C. M. (2001). *Demonstrated Benefits from Social Capital: The Productivity of Farmer Organizations in Gal Oya, Sri Lanka* (SSRN Scholarly Paper ID 252837). Social Science Research Network. https://papers.ssrn.com/abstract=252837

Valentine, G. (2008). Living with difference: Reflections on geographies of encounter. *Progress in Human Geography*, *32*(3), 323–337. https://doi.org/10.1177/0309133308089372

van der Gaag, M. (2002, February 13). *The Resource Generator: Measurement of Individual Social Capital with Concrete Items*. XXII Sunbelt International Social Networks Conference, New Orleans.

van Oorschot, W., & Arts, W. (2005). The social capital of European welfare states: The crowding out hypothesis revisited. *Journal of European Social Policy*, *15*(1), 5–26. https://doi.org/10.1177/0958928705049159

Veenstra, G. (2002). Explicating Social Capital: Trust and Participation in the Civil Space. *The Canadian Journal of Sociology / Cahiers Canadiens de Sociologie*, *27*(4), 547–572. https://doi.org/10.2307/3341590

Venables, W. N., & Ripley, B. D. (2003). *Modern Applied Statistics with S* (4th edition). Springer.

Vezina, M. (2011). *Quality of personal networks: Does living alone make a difference?* (Canadian Social Trends Cagalogue no. 11-008; pp. 60–69). Statistics Canada.

Waldrop, M. M. (1993). *Complexity: The emerging science at the edge of order and chaos*. Simon & Schuster.

Walton, E. (2018). The meaning of community in diverse neighborhoods: Stratification of influence and mental health. *Health & Place*, *50*, 6–15. https://doi.org/10.1016/j.healthplace.2018.01.001

Wang, P., Chen, X., Gong, J., & Jacques-Tiura, A. J. (2014). Reliability and Validity of the Personal Social Capital Scale 16 and Personal Social Capital Scale 8: Two Short Instruments for Survey Studies. *Social Indicators Research*, *119*(2), 1133–1148. https://doi.org/10.1007/s11205-013-0540-3

Warrian, P. (2009, April 30). Biotech in Lunch Buckets: The Curious Knowledge Networks of Steeltown. *ISRN Study on Creative Cities*. IRSN Annual Meeting, Halifax, Nova Scotia. http://sites.utoronto.ca/isrn/publications/NatMeeting/NatSlides/Nat09/papers/WarrianP_Biotech_Lunchbuckets_Paper_%20Conf09.pdf

Warrian, P. (2010). *The Importance of Steel Manufacturing to Canada: A Research Study* (Munk School Briefings, p. 113) [Research Study]. Munk School, University of Toronto. https://munkschool.utoronto.ca/research-articles/the-importance-of-steel-manufacturing-to-canada-a-research-study/

Watts, D. J. (2004). *Six degrees: The science of a connected age*. Norton.

Watts, D. J. (2017). Should social science be more solution-oriented? *Nature Human Behaviour*, *1*(1), 0015. https://doi.org/10.1038/s41562-016-0015

Wedding, L. M., Lecky, J., Gove, J. M., Walecka, H. R., Donovan, M. K., Williams, G. J., Jouffray, J.-B., Crowder, L. B., Erickson, A., Falinski, K., Friedlander, A. M., Kappel, C. V., Kittinger, J. N., McCoy, K., Norström, A., Nyström, M., Oleson, K. L. L., Stamoulis, K. A., White, C., & Selkoe, K. A. (2018). Advancing the integration of spatial data to map human and natural drivers on coral reefs. *PLOS ONE*, *13*(3), e0189792. https://doi.org/10.1371/journal.pone.0189792

Weinberger, D. (2012). *Too Big to Know: Rethinking Knowledge Now That the Facts Aren't the Facts, Experts Are Everywhere, and the Smartest Person in the Room Is the Room*. Basic Books.

Wellman, B., & Rainie, L. (2012). *Networked: The New Social Operating System*. The MIT Press.

Westley, F., Zimmerman, B., & Patton, M. (2006). *Getting to Maybe: How the World Is Changed*. Random House Canada.

Westlund, H., & Adam, F. (2010). Social Capital and Economic Performance: A Meta-analysis of 65 Studies. *European Planning Studies*, *18*(6), 893–919. https://doi.org/10.1080/09654311003701431

Wheatley, M. J. (2006). *Leadership and the New Science: Discovering Order in a Chaotic World* (Third

Edition). Berrett-Koehler Publishers.
White, E. (2010). *Lonely: A Memoir*. Harper.
Whitehead, M., Pennington, A., Orton, L., Nayak, S., Petticrew, M., Sowden, A., & White, M. (2016). How could differences in 'control over destiny' lead to socio-economic inequalities in health? A synthesis of theories and pathways in the living environment. *Health & Place*, *39*, 51–61. https://doi.org/10.1016/j.healthplace.2016.02.002
Wilson, H. F. (2017). On geography and encounter: Bodies, borders, and difference. *Progress in Human Geography*, *41*(4), 451–471. https://doi.org/10.1177/0309132516645958
Wilson, P. (2009, October 26). A Challenged Corner of the City. *Hamilton Spectator*, G.4.
Wilson, R., & Din, A. (2018). Calculating Varying Scales of Clustering Among Locations. *Cityscape; Washington*, *20*(1), 215–231.
Wolfe, P. (2013). Making sense of big data. *Proceedings of the National Academy of Sciences*, *110*(45), 18031–18032.
Wong, D. W. S. (1999). Geostatistics as Measures of Spatial Segregation. *Urban Geography*, *20*(7), 635–647. https://doi.org/10.2747/0272-3638.20.7.635
Woolcock, M. (1998). Social Capital and Economic Development: Toward a Theoretical Synthesis and Policy Framework. *Theory and Society*, *27*(2), 151–208.
Woolcock, M. (2001). The place of social capital in understanding social and economic outcomes. *Canadian Journal of Policy Research*, *2*(1), 11–17.
Woolcock, M., & Narayan, D. (2000). Social Capital: Implications for Development Theory, Research, and Policy. *The World Bank Research Observer*, *15*(2), 225–249.
Yamamoto, Y. (2018). Reconsidering "Desire" and "Style": A Lefebvrian Approach to Democratic Orientation in Planning. *Urban Planning*, *3*(3), 105–112. https://doi.org/10.17645/up.v3i3.1362
Yao, D., van der Hoorn, P., & Litvak, N. (2017). *Average nearest neighbor degrees in scale-free networks*. https://arxiv.org/abs/1704.05707v4
Yasuda, N., Zimmerman, S. I., Hawkes, W., Fredman, L., Hebel, J. R., & Magaziner, J. (1997). Relation of Social Network Characteristics to 5-Year Mortality among Young-Old versus Old-Old White Women in an Urban Community. *American Journal of Epidemiology*, *145*(6), 516–523.
Yeomans, K. A., & Golder, P. A. (1982). The Guttman-Kaiser Criterion as a Predictor of the Number of Common Factors. *Journal of the Royal Statistical Society. Series D (The Statistician)*, *31*(3), 221–229. https://doi.org/10.2307/2987988
Young, F. W., Takane, Y., & de Leeuw, J. (1978). The principal components of mixed measurement level multivariate data: An alternating least squares method with optimal scaling features. *Psychometrika*, *43*(2), 279–281. https://doi.org/10.1007/BF02293871
Zhang, L., Dalyot, S., Eggert, D., & Sester, M. (2011). *Multi-stage Approach to Travel-mode Segmentation and Classification of GPS Traces*.
Zhang, X., Norris, S. L., Gregg, E. W., & Beckles, G. (2007). Social support and mortality among older persons with diabetes. *The Diabetes Educator*, *33*(2), 273–281. https://doi.org/10.1177/0145721707299265
Zhou, W.-X., Sornette, D., Hill, R. A., & Dunbar, R. I. M. (2005). Discrete Hierarchical Organization of Social Group Sizes. *Proceedings: Biological Sciences*, *272*(1561), 439–444.

CPSIA information can be obtained
at www.ICGtesting.com
Printed in the USA
LVHW070729280222
712160LV00005B/115
* 9 7 8 0 9 8 1 2 1 2 0 1 2 *